MONOGRAPHS ON
PHYSICAL BIOCHEMISTRY

GENERAL EDITORS

W. F. HARRINGTON   A. R. PEACOCKE

COOPERATIVE EQUILIBRIA IN PHYSICAL BIOCHEMISTRY

BY
D. POLAND

CLARENDON PRESS · OXFORD

Oxford University Press, Walton Street, Oxford OX2 6DP

OXFORD LONDON GLASGOW NEW YORK
TORONTO MELBOURNE WELLINGTON CAPE TOWN
IBADAN NAIROBI DAR ES SALAAM LUSAKA ADDIS ABABA
KUALA LUMPUR SINGAPORE JAKARTA HONG KONG TOKYO
DELHI BOMBAY CALCUTTA MADRAS KARACHI

ISBN 0 19 854622 X

© Oxford University Press 1978

All rights reserved. No part of this publication may be reproduced, stored in a retrieval system, or transmitted, in any form or by any means, electronic, mechanical, photocopying, recording, or otherwise, without the prior permission of Oxford University Press

Printed in Great Britain by
Thomson Litho Ltd., East Kilbride, Scotland.
Bound by Hunter & Foulis, Edinburgh.

PREFACE

The series of Monographs on Physical Biochemistry is intended to enable advanced undergraduates, graduate students, and research workers starting with no previous experience to achieve a working knowledge of the research-methods in physical biochemistry. In speaking of a research method one usually has an experimental technique in mind. The present book is not about an experimental technique but about a way of interpreting experimental results and of planning new experiments. The basic discipline required for discussing complex co-operative equilibria is statistical mechanics. Yet this book has not been designed primarily as a text in applied statistical mechanics; instead it is intended to meet the requirements outlined in the first sentence above. Thus I start by assuming only that the reader is familiar with the basic concepts of simple chemical equilibrium and at least the rudimentary notions of chemical thermodynamics. The task of taking the reader from this starting point to a working knowledge of present-day research in a short monograph is certainly a challenge. However, I feel it is a challenge well worth taking up. Complex co-operative systems are the rule rather than the exception in biochemistry and the interpretation of these systems should be part of the arsenal of research methods of every experimental physical biochemist, if for no other reason than to obtain an appreciation of the many simplifying assumptions required if one is to make progress in analysing complex systems.

   I have been guided in the present work by my attempts at The Johns Hopkins University to introduce the treatment of cooperative equilibria at many levels in the curriculum, including first-year undergraduate chemistry, introductory physical chemistry, and graduate-level courses in biophysics and statistical mechanics. Thus my reasons for choosing particular material and examples and the order of presentation are largely dictated by students' reactions, positive and negative, over the years at all levels in the curriculum. Almost all the material in this book has been tested by students, and I can only hope it works as well in print as it seems to have done in lectures.

A few words about the organization. The first chapter begins with a discussion of simple chemical equilibrium, with which we assume the reader is thoroughly familiar. Chemical equilibrium is considered as a multiple equilibria between free energy levels with the partition function as a simplifying tool in treating such a free-energy-level system. All the basic equations for treating co-operative equilibria are presented in this chapter as a generalization of simple $A \rightleftharpoons B$ type equilibria. Most of the first chapter can easily be read and understood by well-motivated first-year undergraduates. The next chapter is in a sense a review of the fundamental ideas in thermodynamics, statistical mechanics, and intermolecular forces that are important for a basic understanding of co-operative equilibria. I have included in this chapter those concepts in physical chemistry that are important for the present topic and that I have found to be poorly understood by most students. For example, I find that very few students can give even an order-of-magnitude estimate for the energy of interaction of an ion with a dipole or the entropy of a molecule in the gas phase. Nor do most students have clear notions of the origin of non-ideality effects or their magnitudes. While I do not hold much hope that the properties of biological systems can be accurately calculated from the basic constants of physics, it is very important that the student should know the magnitudes of important interactions and have a sound understanding of their origins. And this is often possible to an accuracy of around 10 per cent using simple models. The purpose of the second chapter, then, is to give the student a more comfortable feeling with regard to those aspects of physical chemistry useful in formulating models for co-operative equilibria.

The rest of the book is concerned with systematic applications of the concepts outlined in the first two chapters. While the applications emphasize conformational transitions in biological macromolecules, the techniques presented are applicable to any co-operative equilibrium. It should be noted that the applications do not represent a systematic coverage of all research on biological macromolecules but rather the examples are chosen to represent important techniques. More detail on many of the topics in the present book can be found in the book *Theory of helix-coil transitions in biopolymers* (Academic Press, 1970) by myself and Professor H.A. Scheraga (this book is referred to in the text as *P&S*). It should be noted, though, that in the years since that book was published many new and improved techniques have been developed. Thus the present volume contains many results and approaches (happily, often involving significant simplifications) that the earlier book does not include.

I wish to thank Professor William Harrington for inviting me to adapt my lectures at The Johns Hopkins University into the form of a short book, and to the Sloan Foundation and the Camille and Henry Dreyfus Foundation for general support during the writing.

September 1974 D.P.
Baltimore, Maryland

CONTENTS

1. SIMPLE MULTIPLE EQUILIBRIA ... 1
   - 1.1. Two-level system ... 1
   - 1.2. Multilevel system ... 5
   - 1.3. Two-level system with degeneracy ... 9
   - 1.4. Multilevel system with degeneracy ... 11
   - 1.5. Simple equilibria between isomers ... 14
   - 1.6. Simple ionization equilibria ... 18
   - 1.7. Examples of proton association ... 22
   - 1.8. Equilibria between free energy levels ... 33
   - 1.9. Physical cluster model ... 34
   - 1.10. Some examples of binding equilibria ... 50
   - 1.11. Summary ... 54
   - Notes and References ... 55

2. BACKGROUND: PARTITION FUNCTIONS AND INTERMOLECULAR FORCES ... 56
   - 2.1. Partition functions ... 57
     - 2.1.1. Thermodynamics ... 58
     - 2.1.2. Statistical mechanics: the molecular interpretation of thermodynamics ... 72
     - 2.1.3. Einstein crystal ... 77
     - 2.1.4. Ideal Gas ... 80
     - 2.1.5. A gas-to-crystal model; maximum-term approximation ... 83
     - 2.1.6. Grand partition function ... 88
     - 2.1.7. Physical cluster model from the grand partition function ... 91
     - 2.1.8. Molecule grand partition function ξ ... 96
     - 2.1.9. Classical statistical mechanics ... 99
     - 2.1.10. Dilute solutions ... 102
   - 2.2. Intermolecular forces ... 103
     - 2.2.1. Electrostatic energy ... 103
     - 2.2.2. Polarizability of molecules ... 113
     - 2.2.3. Statistical potentials ... 116
     - 2.2.4. van der Waals interactions ... 117
     - 2.2.5. Repulsive potentials ... 119
     - 2.2.6. Simple net potentials ... 119
     - 2.2.7. Internal rotation potentials ... 124
     - 2.2.8. The hydrogen bond ... 236
     - 2.2.9. Empirical molecular potentials ... 127
     - 2.2.10. Bond bending and stretching ... 129
     - 2.2.11. Coulombic interactions in solution ... 130
     - 2.2.12. Hydrogen bonds in polar solvents ... 132
     - 2.2.13. Hydrophobic bonds ... 133
   - 2.3. Simple applications ... 134
     - 2.3.1. Crystal potential and sublimation energies ... 134
     - 2.3.2. Average distances between molecules ... 140
     - 2.3.3. Thermodynamics of solutions of salts ... 142
     - 2.3.4. Chemical potential for dilute salt solutions ... 144
     - 2.3.5. Excluded volume ... 152
     - 2.3.6. Non-ideality from association ... 160
     - 2.3.7. Thermodynamics of argon (solid-liquid-gas) ... 164

		2.3.8. Thermodynamics of water (liquid)	173
		2.3.9. Dimerizations	176
	2.4.	Summary	178
	Notes and References		179

3. ASSOCIATION OF BIOPOLYMERS — 181

- 3.1. Dimerization of oligonucleotides — 181
  - 3.1.1. Basic equilibrium — 182
  - 3.1.2. Concentration dependence of the equilibrium — 183
  - 3.1.3. Temperature dependence of the equilibrium — 187
  - 3.1.4. Partition function for the double-strand complex — 188
  - 3.1.5. Example of oligo(A) — 195
- 3.2. Poly(A–T) — 200
  - 3.2.1. Dimerization of short chains — 200
  - 3.2.2. Properties of loops — 203
  - 3.2.3. Statistical weights for nucleic acids — 206
- 3.3. Association of (Gly-Pro-Pro)$_N$ — 210
  - 3.3.1. Triple-helix equilibrium — 210
  - 3.3.2. Thermodynamic parameters — 213
  - 3.3.3. Properties of the equilibrium — 215
- 3.4. Statistical weights for natural collagen — 217
  - 3.4.1. Stability of collagen as a function of amino-acid composition — 218
  - 3.4.2. Simple model for the melting temperature — 220
- 3.5. Sequence statistics in natural collagen — 222
- 3.6. Summary — 225
- Notes and References — 227

4. LINEAR CHAIN PARTITION FUNCTIONS: TECHNIQUE AND APPLICATIONS — 228

- 4.1. Matrix method — 228
- 4.2. Sequence conditional probabilities — 237
- 4.3. Combinatorial formulation — 244
- 4.4. Simple model for polyethylene — 246
- 4.5. α-helix in polypeptides — 252
- 4.6. Long-chain synthetic polynucleotides — 266
- 4.7. Evaluation of derivatives of partition functions — 271
- Notes and References — 273

5. COOPERATIVE BINDING TO BIOPOLYMERS — 275

- 5.1. Titration of rigid macromolecules — 275
- 5.2. Binding of Mg^{2+} to DNA — 283
- 5.3. Cooperative binding to surfaces — 286
- 5.4. Cooperative binding of oligomers to polymers — 290
- 5.5. Summary — 301
- Notes and References — 303

6. COOPERATIVE EQUILIBRIA IN SPECIFIC-SEQUENCE MACROMOLECULES — 304

- 6.1. Probability profiles; nearest-neighbour interactions — 305
- 6.2. Probability profiles; long-range correlations — 311
- 6.3. Nature of the equilibrium in specific-sequence molecules — 321
- 6.4. Probability of interior unwinding in collagen — 325
- 6.5. Probability profiles for globular proteins — 327
- 6.6. Summary — 340
- Notes and References — 341

APPENDIX: A SHORT REVIEW OF MATRIX ALGEBRA — 342

1
SIMPLE MULTIPLE EQUILIBRIA

We shall begin by considering a simple general method for treating multiple equilibria; to do this we start by discussing the simplest possible equilibrium.

1.1. Two-level system

The simplest equilibrium occurs when a system (hypothetical molecule) has only two quantum-mechanical energy levels available to it, as illustrated in Fig. 1.1(a). For such a system the ratio of the occupation probabilities for the two levels is given by

$$p_2/p_1 = \exp(-\Delta\varepsilon/RT), \qquad (1.1)$$

$\Delta\varepsilon$ being the difference in energy between the two levels, R the gas constant, and T the temperature in degrees kelvin. Eqn (1.1), the Boltzmann distribution, will be our fundamental starting point; the derivation of eqn (1.1) can be found in any text on physical chemistry or statistical mechanics.[1] Of more

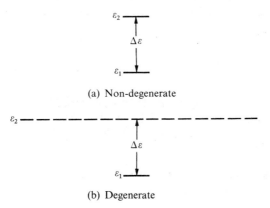

FIG. 1.1. Simple two-level energy schemes: (a) non-degenerate; (b) two-level scheme with upper-level degenerate.

interest than the ratio of the occupation probabilities is an explicit relation for each of the occupation probabilities. Using the definition of $\Delta\varepsilon$ and the conservation of probabilities,

$$p_1 + p_2 = 1,$$

$$\Delta\varepsilon = \varepsilon_2 - \varepsilon_1$$

(1.2)

we easily obtain

$$p_1 = \frac{\exp(-\varepsilon_1/RT)}{\exp(-\varepsilon_1/RT) + \exp(-\varepsilon_2/RT)} = \frac{1}{1 + \exp(-\Delta\varepsilon/RT)}$$

$$p_2 = \frac{\exp(-\varepsilon_2/RT)}{\exp(-\varepsilon_1/RT) + \exp(-\varepsilon_2/RT)} = \frac{\exp(-\Delta\varepsilon/RT)}{1 + \exp(-\Delta\varepsilon/RT)}$$

(1.3)

Eqn (1.3) illustrates the fact that it is only the difference in energy that enters into the expressions for the occupation probabilities although they can be written explicitly in terms of ε_1 and ε_2. Since analogues of eqn (1.3) will play a central role in our discussion of complex equilibria, it is essential to be thoroughly familiar with how p_1 and p_2 vary as a function of $\Delta\varepsilon$ and T.

The main ingredient in eqn (1.3) is, of course, $\exp(-\Delta\varepsilon/RT)$. The temperature dependence of this factor is most easily demonstrated by examining the extremes of temperature. As the temperature goes to absolute zero we have

$$e^{-\Delta\varepsilon/0} = e^{-\infty} = 1/e^{\infty} = 0,$$

while in the limit of very large temperature we find

$$e^{-\Delta\varepsilon/\infty} = e^{-0} = 1.$$

Thus $\exp(-\Delta\varepsilon/RT)$ varies from zero to unity as the temperature is increased

T	$\exp(-\Delta\varepsilon/RT)$
0	0
∞	1

(1.4)

SIMPLE MULTIPLE EQUILIBRIA

From the behaviour outlined in expression (1.4) we can immediately evaluate the limiting behaviour of p_1 and p_2 as given in eqn (1.3)

T	p_1	p_2
0	1	0
∞	$\frac{1}{2}$	$\frac{1}{2}$.

The above tabulation illustrates several general features of the behaviour of occupation probabilities as a function of temperature: at 0 K the probability of being in the lowest energy level is unity, the occupation probabilities of the higher levels being zero; and, as the temperature is increased, the occupation probabilities tend to become equal.

Having examined the limiting behaviour at the extremes of temperature it is useful to observe the temperature range over which p_1 and p_2 make the transition from the low- to the high-temperature limit. We can estimate this temperature by solving eqn (1.3) for the temperature at which $p_1 = \frac{3}{4}$ (which is half way between the low and high temperature limits). We obtain $RT = \Delta\varepsilon/\ln 3$ or, since $\ln 3 \sim \ln e \simeq 1$, $RT \simeq \Delta\varepsilon$. The estimate of $RT \simeq \Delta\varepsilon$ for the temperature at which the upper levels start to become significantly populated is a general one for any energy-level scheme and simply states that, when RT is of the same order of magnitude as the spacings between levels, we find the upper levels beginning to be populated.

Since RT plays such an important role in understanding the population of levels, the value of this factor at common temperatures should be familiar. In the units of $J\ K^{-1}\ mol^{-1}$, R has the value

$$R = 8.31\ J\ K^{-1}\ mol^{-1}. \tag{1.5}$$

Throughout this book for numerical estimates we will use the rounded value of $R = 8.3$. The factor RT at several temperatures is then

°C	K	RT J mol^{-1}
−100	173	1435
0	273	2265
25	298	2475
100	373	3095

Notice that RT is given in J mol^{-1} not kJ mol^{-1}. Hence, at 298 K $RT \sim$ 2500 J mol^{-1} = 2.5 kJ mol^{-1}; in evaluating $\Delta\varepsilon/RT$ we must be careful to express both $\Delta\varepsilon$ and RT in the same units. Conversion factors to other common energy units that are useful to remember are

1 electronvolt per molecule = 96.5 kJ mol^{-1}

and   4.184 J mol^{-1} = 1 cal mol^{-1}.

We will use units of kJ mol^{-1} throughout this book. In most books on physical chemistry and statistical mechanics one finds the Boltzmann expression, eqn (1.1), written as $\exp(-\Delta\varepsilon/kT)$; the difference between using k and R is simply the units used to express $\Delta\varepsilon$, k being used when one is treating ε as the energy per molecule, R when one is referring to the energy per mole.

Returning now to the question of how p_1 and p_2 of eqn (1.3) vary as a function of $\Delta\varepsilon$, Table 1.1 shows these quantities as a function of $\Delta\varepsilon$ and temperature. The most important result of

TABLE 1.1

Occupation probabilities for the two-level system as a function of $\Delta\varepsilon/RT$ (the values of $\Delta\varepsilon$ are for T = 300 K)

$\Delta\varepsilon/RT$	p_1	p_2	$\Delta\varepsilon$ (kJ mol^{-1})
0	0.50	0.50	0.00
$\frac{1}{4}$	0.55	0.45	0.63
$\frac{1}{2}$	0.62	0.38	1.25
1	0.73	0.27	2.51
2	0.88	0.12	5.02
4	0.98	0.02	10.03
∞	1.00	0.00	∞

SIMPLE MULTIPLE EQUILIBRIA 5

Table 1.1 is the fact that for $\Delta\varepsilon$ greater than about 2.5 kJ mol^{-1} the system is still predominantly in the ground state (lowest energy level) at room temperature. Thus it does not take a very large energy spacing to essentially prohibit occupation of higher levels.

1.2. Multilevel system

One artificial feature of the two-level system is just that it consists of only two levels. Since most real systems have in fact an infinite number of energy levels available, the question arises as to how general are the features that we have just discussed. For example, we might wonder whether the presence of an infinite number of levels above the ground state, even though each has a very small occupation probability, might put a significant drain on the occupation probability of the lowest level.

The simplest extension of the two-level system is the case where we have an infinite number of levels with the same spacing $\Delta\varepsilon$ between successive levels as shown in Fig. 1.2(a). Although this is the quantum-mechanical energy-level scheme for

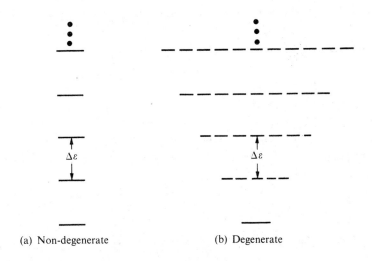

FIG. 1.2. Simple multilevel energy schemes: (a) non-degenerate, evenly spaced levels; (b) degenerate, evenly spaced levels.

the one-dimensional harmonic oscillator, we use it here simply to illustrate equilibrium for a simple multilevel system. It is convenient for the sake of future notation to label the lowest level as the zeroth. Extending eqn (1.1) to this system gives for any two adjacent levels

$$p_1/p_0 = \exp(-\Delta\varepsilon/RT),$$

$$p_2/p_1 = \exp(-\Delta\varepsilon/RT),\quad (1.6)$$

etc.

If we multiply p_1/p_0 times p_2/p_1, we obtain

$$p_2/p_0 = \exp(-2\Delta\varepsilon/RT), \quad (1.7)$$

which is easily generalized to give

$$p_n/p_0 = \exp(-n\Delta\varepsilon/RT). \quad (1.8)$$

It will be a recurrent theme in this book that the ratio as given in eqn (1.8) is much more useful than the ratios given in eqn (1.6), the reason being simply that eqn (1.8) is given in terms of p_n for any level and only one other level, p_0. Explicitly, we can solve eqn (1.8) for the p_n:

$$p_n = p_0 \exp(-n\Delta\varepsilon/RT). \quad (1.9)$$

If we know p_0 (and of course $\Delta\varepsilon$) then eqn (1.9) immediately gives all the other p_n. Since the probability of being in any of the levels is unity (conservation of occupation probability) we have

$$\sum_{n=0}^{\infty} p_n = 1. \quad (1.10)$$

Using eqn (1.9) for the p_n we then obtain an equation for p_0 alone:

SIMPLE MULTIPLE EQUILIBRIA

$$p_0 \sum_{n=0}^{\infty} \exp(-n\Delta\epsilon/RT) = 1$$

or

$$p_0 = 1 / \sum_{n=0}^{\infty} \exp(-n\Delta\epsilon/RT). \tag{1.11}$$

The p_n of eqn (1.9) are then given explicitly as

$$p_n = \frac{\exp(-n\Delta\epsilon/RT)}{\sum_{n=0}^{\infty} \exp(-n\Delta\epsilon/RT)} \tag{1.12}$$

The sum in eqn (1.11), which we have arrived at in a specific example, is of general significance in statistical mechanics. In general, for an energy-level scheme designated by ϵ_n (where the levels are not necessarily equally spaced) we have

$$p_n = \exp(-\epsilon_n/RT)/q,$$
$$q = \sum \exp(-\epsilon_n/RT). \tag{1.13}$$

The quantity q in eqn (1.13) is known as the partition function, since in eqn (1.13) it acts as a normalizing factor allowing us to calculate how the probabilities of occupation are divided up or partitioned out among the levels.

For the example at hand, a closed form for q can be easily obtained. Letting

$$x = \exp(-\Delta\epsilon/RT), \tag{1.14}$$

then

$$q = \sum_{n=0}^{\infty} x^n = \frac{1}{1-x}, \tag{1.15}$$

the sum in eqn (1.15) being simply the geometric series. Eqn (1.12) is now given in closed form as

$$p_0 = (1-x),$$
$$p_n = (1-x)x^n. \tag{1.16}$$

Since $x = \exp(-\Delta\varepsilon/RT)$ is the same parameter that occurred in eqn (1.3) for the two-level problem, it is instructive to calculate the probability of being in the ground state for the present multilevel model for the same values of $\Delta\varepsilon$ and T as used in Table 1.1 to see explicitly how the presence of an infinite number of upper levels affects the equilibrium. This calculation is shown in Table 1.2, where we see that indeed the presence of an infinite number of upper levels does put a drain on the occupation probability of the ground state, but that this becomes an increasingly small effect for $\Delta\varepsilon$ greater than about RT. It is still true that $RT \simeq \Delta\varepsilon$ gives the temperature at which the upper levels begin to be populated significantly.

TABLE 1.2.

Ground-state probabilities for the two-level system with the upper level degeneracies $\omega = 1$ and $\omega = 10$ (the values of $\Delta\varepsilon$ are for T = 300 K)

$\Delta\varepsilon/RT$	$p_1(\omega = 1)$	$p_1(\omega = 10)$	$\Delta\varepsilon$(kJ mol^{-1})
0	0.50	0.09	0.00
$\frac{1}{4}$	0.55	0.10	0.63
$\frac{1}{2}$	0.62	0.14	1.25
1	0.73	0.21	2.51
2	0.88	0.42	5.02
4	0.98	0.85	10.03
∞	1.00	1.00	∞

One feature of the multilevel system that is qualitatively different from the two-level system is the nature of the probability distribution at high temperature. For the two-level system $p_1 = p_2 = \frac{1}{2}$ as $RT \gg \Delta\varepsilon$. For the multilevel system the probability of all the levels tends to be equalized at high temperature, but since there are an infinite number of levels, the probabilities all become small at high temperature. Fig. 1.3 shows $p_n(T)$ at three temperatures: $RT = \frac{1}{2}\Delta\varepsilon$, $RT = \Delta\varepsilon$, and $RT = 2\Delta\varepsilon$. Since from expression (1.4) $x (= \exp(-\Delta\varepsilon/RT))$ is always less than unity, $(1-x)x^n$ is a monotonically decreasing function of x, the ground state always being the most probable

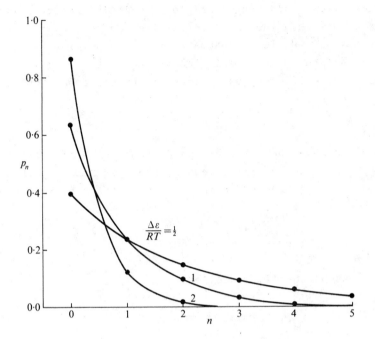

FIG. 1.3. Occupation probability distribution for the energy-level scheme of Fig. 1.2(a). The curves shown are for $\Delta\varepsilon/RT = \frac{1}{2}$, 1, and 2 respectively, the distribution becoming broader for smaller $\Delta\varepsilon/RT$ (or higher temperature).

state. The kind of behaviour shown in Fig. 1.3, i.e. the tendency for the occupation probability distribution to become broader as the temperature is increased, is a general one although, of course, the fine details of the distribution depend on the nature of the spacing in the energy-level scheme. Fig. 1.3 serves as a fundamental starting point for developing intuition and physical insight into the nature of multiple equilibria.

1.3. Two-level system with degeneracy

Although it would be unnecessarily tedious to discuss in detail all the many simple energy-level schemes that readily come to mind, there is a modification of each of the two energy-level schemes we have discussed that is of fundamental importance in understanding multiple equilibria. And that modification occurs when the upper levels are degenerate. Fig. 1.1(b) shows the two-level scheme where the upper level is ω-fold degenerate. If we label each of the states of the upper level as a, b, etc. then eqn (1.1) still holds, for example,

$$p_{2a}/p_1 = \exp(-\Delta\varepsilon/RT). \qquad (1.17)$$

If we want to know the ratio of the probabilities of the system being in the second and first levels, where by being in the second level we mean being in any of the ω states of the second level, then we have

$$p_2/p_1 = \omega \exp(-\Delta\varepsilon/RT). \qquad (1.18)$$

The analogues of eqn (1.3) are then

$$p_1 = \frac{1}{1 + \omega \exp(-\Delta\varepsilon/RT)},$$

$$p_2 = \frac{\omega \exp(-\Delta\varepsilon/RT)}{1 + \omega \exp(-\Delta\varepsilon/RT)}. \qquad (1.19)$$

In the limits of low and high temperature p_1 and p_2 take on the values

T	p_1	p_2
0	1	$1/(1+\omega)$
∞	0	$\omega/(1+\omega)$

(1.20)

When $\omega = 1$, which is the simple two-level system discussed previously, $p_1 = p_2 = \frac{1}{2}$ at high temperature. As ω becomes significantly greater than unity, then p_2 approaches unity and p_1 approaches zero at high temperature. Another effect of degeneracy is to make the transition to the upper level much sharper, as is shown in Fig. 1.4, where $p_1(T)$ and $p_2(T)$ are plotted for $\Delta\varepsilon = 2.5$ kJ mol^{-1} and $\omega = 10$ and compared with the behaviour of the simple two-level system (same $\Delta\varepsilon$, $\omega = 1$).

SIMPLE MULTIPLE EQUILIBRIA

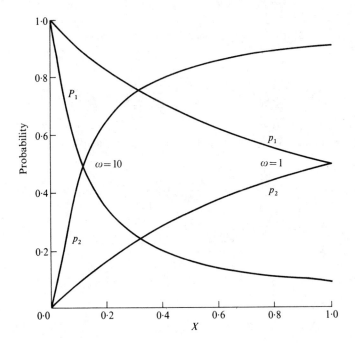

FIG. 1.4. Occupation probability distribution for the two-level energy schemes of Fig. 1.1.

1.4. Multilevel system with degeneracy

We have seen that degeneracy of the upper level can produce marked changes in how p_1 and p_2 vary with temperature for the two-level system. To study the influence of degeneracy in the multilevel system we introduce the simple modification that the nth level is $(n+1)$-fold degenerate, as is shown in Fig. 1.2(b), again retaining the feature of equally spaced levels. Then the analogue of eqn (1.12) for the occupation probability of the nth level is

$$p_n = \frac{(n+1)\exp(-n\Delta\varepsilon/RT)}{\sum_{n=0}^{\infty}(n+1)\exp(-n\Delta\varepsilon/RT)} . \qquad (1.21)$$

Using the definition of x given in eqn (1.14) the partition function q is

$$q = \sum_{n=0}^{\infty} (n+1) \exp(-n\Delta\varepsilon/RT) \qquad (1.22)$$

$$= \sum_{n=0}^{\infty} n\, x^n + \sum_{n=0}^{\infty} x^n .$$

In eqn (1.15) we have already evaluated $\Sigma\, x^n$. The quantity $\Sigma\, nx^n$ is also readily evaluated by observing that

$$\frac{d\, \Sigma\, x^n}{dx} = \Sigma\, n x^{n-1}$$

Hence

$$\Sigma\, n\, x^n = x\, \frac{d\, \Sigma\, x^n}{dx} = x\, \frac{d\{1/(1-x)\}}{dx}$$

or

$$\sum_{n=1}^{\infty} n\, x^n = \frac{x}{(1-x)^2} . \qquad (1.23)$$

Then

$$q = x/(1-x)^2 + 1/(1-x) = 1/(1-x)^2$$

and p_n of eqn (1.21) is given explicitly as

$$p_n = (1-x)^2 (n+1) x^n . \qquad (1.24)$$

Once again (see expression (1.4)) x is less than unity and, hence, x^n is a monotonically decreasing function of x. Thus p_n, which varies as $(n+1)x^n$, is the product of one function, $(n+1)$, that is monotonically increasing with another, x^n, that is monotonically decreasing; the result is a distribution that has a maximum at an intermediate value of n for x sufficiently close to unity.

Fig. 1.5 gives $p_n(T)$ for the present case with degeneracy for the same values of x (hence $\Delta\varepsilon$) as the previous calculation (Fig. 1.3) for the multilevel system without degeneracy. We see that the introduction of degeneracy in the upper levels increases the occupation probabilities of these levels, this being

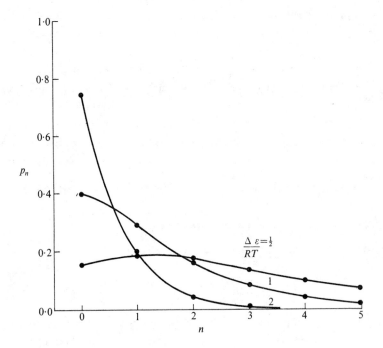

FIG. 1.5. Occupation probability distribution for the energy-level scheme of Fig. 1.2(b). The curves shown are for $\Delta\varepsilon/RT = \frac{1}{2}$, 1, and 2 respectively, the distribution becoming broader for smaller $\Delta\varepsilon/RT$ (or higher temperature).

due simply to the fact that there are more possibilities (states) at the higher energy levels.

It is instructive at this point to summarize one feature of the four models we have discussed, namely the probability of being in the ground state, which is given below (with $x = \exp(-\Delta\varepsilon/RT)$).

	Ground-state probability	
	non-degenerate	degenerate
two-level	$1/(1+x)$	$1/(1+\omega x)$
multilevel	$(1-x)$	$(1-x)^2$

The reason that the probability of being in the ground state is an important quantity to understand is that if the probability of the ground state is very close to unity, then we need not be concerned about multiple equilibria at all. An easy number to remember is the value of x when $\Delta \varepsilon = RT$ since then $x = e^{-1} \simeq \frac{1}{3}$ (also, for $\Delta \varepsilon = 2RT$, $x \sim \frac{1}{10}$).

While the models we have been discussing are idealized quantum-mechanical energy-level systems, we now will show that the mathematics involved for treating simple multiple equilibria in real systems is exactly the same as for the simple systems just discussed, the only change being new symbols. And most important, the physical nature of the occupation probability distributions is the same.

1.5. Simple equilibria between isomers

Consider the simplest chemical reaction

$$A \rightleftharpoons B \tag{1.25}$$

with equilibrium constant K. The reaction given in eqn (1.25) involves no gain or loss of chemical species, the simplest examples of such reactions being isomerizations, for example, between different conformations of a molecule such as the interconversion between the chair and boat forms of cyclohexane or the *cis* and *trans* forms of 1,2-dichloroethylene. The law of mass action tells us that the concentrations of A and B are related by

$$[B]/[A] = K. \tag{1.26}$$

In eqn (1.25) the total number of molecules present is conserved which we may write as

$$[A] + [B] = [A]_0 + [B]_0 = c_0, \tag{1.27}$$

where $[A]_0$ and $[B]_0$ are the initial concentrations of A and B respectively, the sum of which we denote by c_0. Introducing the mole fractions

$$\begin{aligned} f_A &= [A]/c_0, \\ f_B &= [B]/c_0, \end{aligned} \tag{1.28}$$

SIMPLE MULTIPLE EQUILIBRIA

eqn (1.26) becomes

$$f_B/f_A = K. \tag{1.29}$$

Since by definition

$$f_A + f_B = 1, \tag{1.30}$$

we can eliminate either f_A or f_B from eqn (1.29) and solve for the remaining variable. We easily obtain

$$f_A = \frac{1}{1+K},$$
$$f_B = \frac{K}{1+K}. \tag{1.31}$$

From thermodynamics we have the relation[2]

$$K = \exp(-\Delta g^0/RT), \tag{1.32}$$

where

$$\Delta g^0 = g_B^0 - g_A^0. \tag{1.33}$$

The quantity g^0 is the standard Gibbs free energy per mole for each substance; this is simply a number having the units of energy per mole characteristic of each species.

Using eqn (1.32) in eqn (1.31) we have

$$f_A = \frac{1}{1 + \exp(-\Delta g^0/RT)} = \frac{\exp(-g_A^0/RT)}{\exp(-g_A^0/RT) + \exp(-g_B^0/RT)},$$

$$f_B = \frac{\exp(-\Delta g^0/RT)}{1 + \exp(-\Delta g^0/RT)} = \frac{\exp(-g_B^0/RT)}{\exp(-g_A^0/RT) + \exp(-g_B^0/RT)}. \tag{1.34}$$

On comparing eqn (1.34) with eqn (1.3) for the simple two-level system, we see that they are exactly the same with Δg^0 replacing $\Delta \epsilon$. Hence, immediately we can say that, if Δg^0 is greater than about RT, the probability of B is very small, species A playing the role of the ground state or lowest energy state (except that in the present case we are not talking simply about energy but about free energy).

Since the mole fractions are playing exactly the same role as the occupation probabilities in our discussion of simple energy-level schemes, we will drop the mole fraction notation and adopt the notion of the probability of a particular chemical species, in this case p_A and p_B.

Introducing a third species

$$\begin{aligned} A &\rightleftharpoons B \quad K_{AB}, \\ A &\rightleftharpoons C \quad K_{AC}, \end{aligned} \tag{1.35}$$

where

$$\begin{aligned} K_{AB} &= \exp(-\Delta g^0_{AB}/RT), & K_{AC} &= \exp(-\Delta g^0_{AC}/RT), \\ \Delta g^0_{AB} &= g^0_B - g^0_A, & \Delta g^0_{AC} &= g^0_C - g^0_A, \end{aligned} \tag{1.36}$$

with species probabilities defined as

$$p_A = [A]/c_0, \quad p_B = [B]/c_0, \quad p_C = [C]/c_0$$

where c_0 is the total concentration of A, B, and C, we find in analogy with eqn (1.34),

$$p_A = \frac{\exp(-g^0_A/RT)}{\exp(-g^0_A/RT) + \exp(-g^0_B/RT) + \exp(-g^0_C/RT)} = \frac{1}{1+K_{AB}+K_{AC}},$$

$$p_B = \frac{\exp(-g^0_B/RT)}{\exp(-g^0_A/RT) + \exp(-g^0_B/RT) + \exp(-g^0_C/RT)} = \frac{K_{AB}}{1+K_{AB}+K_{AC}},$$

$$p_C = \frac{\exp(-g^0_C/RT)}{\exp(-g^0_A/RT) + \exp(-g^0_B/RT) + \exp(-g^0_C/RT)} = \frac{K_{AC}}{1+K_{AB}+K_{AC}}.$$

$$\tag{1.37}$$

The generalization to N species is simple. For the equilibria

SIMPLE MULTIPLE EQUILIBRIA

$$A_1 \rightleftharpoons A_2 \rightleftharpoons A_3 \rightleftharpoons \ldots \rightleftharpoons A_n \rightleftharpoons \ldots \rightleftharpoons A_{N-1} \rightleftharpoons A_N \qquad (1.38)$$

with equilibrium constants K_{12}, K_{23}, etc. we have

$$p_n = \frac{\exp(-g_n^0/RT)}{\sum_{n=1}^{N} \exp(-g_n^0/RT)}, \qquad (1.39)$$

where the sum over $\exp(-g_n^0/RT)$ plays the role of the partition function (see eqn (1.13)). Note that if, instead of eqn (1.38), the reactions are written as

$$\begin{aligned} A_1 &\rightleftharpoons A_2, & K_{12}, \\ A_1 &\rightleftharpoons A_3, & K_{13}, \\ &\ldots \\ A_1 &\rightleftharpoons A_n, & K_{1n}, \\ &\ldots \\ A_1 &\rightleftharpoons A_N, & K_{1N}, \end{aligned} \qquad (1.40)$$

then eqn (1.39) becomes

$$p_n = \frac{K_{1n}}{\sum K_{1n}}, \qquad (1.41)$$

where $K_{11} = 1$. Thus writing all the equilibria with reference to a single species (here A_1) leads to great simplification in expressing the probability distribution. Of course, one set of equilibrium constants can be expressed in terms of another by appropriately adding or subtracting reactions, e.g.

$$\begin{aligned} & A_1 \rightleftharpoons A_2, & K_{12} \\ + \; & A_2 \rightleftharpoons A_3, & K_{23} \\ \hline & A_1 \rightleftharpoons A_3, & K_{13} = K_{12}K_{23} \end{aligned} \qquad (1.42)$$

The significance of eqn (1.39) is that equilibria between N isomers is mathematically exactly analogous to the probability distribution in an N-level simple quantum-mechanical energy-level scheme: there are N free energy levels g_n^0, the probability of a given free energy level being given by $\exp(-g_n^0/RT)$ divided by the sum of all such terms. The difference between equilibrium between energy levels ε_n and free energy levels g_n^0 is that the free energy is itself a function of temperature. Before studying this important difference, however, we wish to pursue the analogy of equilibrium between energy levels with the equilibrium between free energy levels for chemical reactions that involve changes in concentration.

1.6. Simple ionization equilibria

The dissociation of a proton in a simple organic acid is described by the reaction

$$AH \rightleftharpoons A^- + H^+, \quad K_d. \tag{1.43}$$

First we turn the reaction around to represent the association of a proton

$$A^- + H^+ \rightleftharpoons AH, \quad K_a = K_d^{-1}. \tag{1.44}$$

The reason for doing this will soon become apparent. The law of mass action[3] gives

$$\frac{[AH]}{[A^-][H^+]} = K_a. \tag{1.45}$$

With the relations

$$[AH] + [A^-] = c_0,$$
$$[A^-] = [H^+], \tag{1.46}$$

we can solve eqn (1.45) for all the different species involved (in addition we could introduce the hydrolysis of water as a separate source of protons). In practice we measure the concentration of protons with a pH meter; hence, we take [H+] as a variable that is experimentally available (solving eqns (1.45) and (1.46) never works very well since other sources of acids, e.g. CO_2, will effect the concentration of protons; if we

SIMPLE MULTIPLE EQUILIBRIA

measure $[H^+]$ then eqn (1.45) gives excellent results at dilute concentrations).

Treating $[H^+]$ as known we can rearrange eqn (1.45) to read

$$\frac{[AH]}{[A^-]} = [H^+]K_a. \tag{1.47}$$

Introducing species probabilities (or mole fractions)

$$p_{AH} = [AH]/c_0,$$

$$p_{A^-} = [A^-]/c_0, \tag{1.48}$$

$$p_{AH} + p_{A^-} = 1,$$

eqn (1.47) becomes

$$\frac{p_{AH}}{p_{A^-}} = [H^+]K_a, \tag{1.49}$$

which is analogous to eqn (1.29), which is analogous to eqn (1.1). The solution of eqn (1.49) using eqn (1.48) is

$$p_{A^-} = \frac{1}{1 + [H^+]K_a},$$

$$p_{AH} = \frac{[H^+]K_a}{1 + [H^+]K_a}, \tag{1.50}$$

which is the analogue of eqn (1.31), itself the analogue of eqn (1.3).

A dibasic acid has the two association reactions

$$A^{2-} + H^+ \rightleftharpoons AH^-, \quad K_1,$$

$$A^{2-} + 2H^+ \rightleftharpoons AH_2, \quad K_2. \tag{1.51}$$

Again the law of mass action gives

$$[AH^-]/[A^{2-}] = [H^+]K_1,$$

$$[AH_2]/[A^{2-}] = [H^+]^2 K_2. \tag{1.52}$$

With the introduction of the species probabilities eqn (1.52) gives

$$p_{AH^-} = p_{A^{2-}} [H^+] K_1,$$
$$p_{AH_2} = p_{A^{2-}} [H^+]^2 K_2, \qquad (1.53)$$

which upon using the conservation relation

$$p_{A^{2-}} + p_{AH^-} + p_{AH_2} = 1 \qquad (1.54)$$

yields

$$p_{A^{2-}} (1 + [H^+] K_1 + [H^+]^2 K_2) = 1 \qquad (1.55)$$

or

$$p_{A^{2-}} = 1/\xi, \quad p_{AH^-} = [H^+] K_1/\xi,$$
$$p_{AH_2} = [H^+]^2 K_2/\xi, \qquad (1.56)$$

where the quantity[4]

$$\xi = 1 + [H^+] K_1 + [H^+]^2 K_2 \qquad (1.57)$$

plays the role of the partition function q in eqn (1.13).

As with previous equilibria, the generalization to N species is straightforward. We have the associations

$$\begin{aligned} A_0 &\rightleftharpoons A_0, & K_0 &= 1, \\ A_0 + H^+ &\rightleftharpoons A_1, & K_1, & \\ A_0 + 2H^+ &\rightleftharpoons A_2, & K_2, & \\ &\cdots & & \\ A_0 + nH^+ &\rightleftharpoons A_n, & K_n, & \\ &\cdots & & \\ A_0 + NH^+ &\rightleftharpoons A_N, & K_N, & \end{aligned} \qquad (1.58)$$

where for simplicity the subscript indicates the number of

SIMPLE MULTIPLE EQUILIBRIA

bound protons. Then

$$p_{A_n} = p_{A_0} [H^+]^n K_n \tag{1.59}$$

with

$$\sum_{n=0}^{N} p_{A_n} = 1 \tag{1.60}$$

gives

$$p_{A_0} \sum_{n=0}^{N} [H^+]^n K_n = 1. \tag{1.61}$$

The utility of eqns (1.58)–(1.61) is that we obtain one equation in one unknown (eqn (1.61)) in terms of which all the other unknowns are given (eqn (1.59)). Thus

$$p_{A_n} = [H^+]^n K_n / \xi,$$

$$\xi = \sum_{n=0}^{N} [H^+]^n K_n. \tag{1.62}$$

The average number of protons bound is given simply in terms of the p_{A_n} as

$$\langle n \rangle = \sum_{n=0}^{N} n p_{A_n}$$

$$= \frac{\Sigma n [H^+]^n K_n}{\Sigma [H^+]^n K_n}. \tag{1.63}$$

We observe that the $\langle n \rangle$ of eqn (1.63) is also given simply as the appropriate derivative of ξ, namely,

$$\langle n \rangle = \frac{[H^+]}{\xi} \frac{\partial \xi}{\partial [H^+]} = \frac{\partial \ln \xi}{\partial \ln [H^+]}. \tag{1.64}$$

We will use analogues of eqn (1.64) throughout this book where an average quantity is given as the appropriate logarithmic derivative of the partition function.

Note that eqns (1.62) and (1.63) are general relations for the distribution of species and the average number of protons bound for any molecule with N dissociable protons (in solutions dilute enough so that there are no significant interactions among the various species). Of course these equations involve N constants K_n which for complex molecules like proteins are not known *a priori*. Thus the significant problem remaining is that of reducing the whole set of K_n to a smaller set of parameters, a problem we will take up later in this book.

We turn now to giving some simple examples for the formalism just developed using as examples simple di- and tribasic acids of importance to biological systems.

1.7. Examples of proton association

Since a great many molecules of biological importance can bind or dissociate protons, the examples given here will serve not only to illustrate the technique of treating multiple dissociation but also give some insight into the state of charge of important groups at various pHs.

For the dissociation of a single proton as described in eqn (1.43), the dissociation constant K_d is commonly written as

$$K_d = 10^{-pK}, \quad (1.65)$$

the exponent pK signifying the negative power of 10 for K. Our equations are much simpler when written in terms of the reverse reaction, for which

$$K_a = 10^{+pK}. \quad (1.66)$$

Using the definition of pH

$$[H^+] = 10^{-pH}, \quad (1.67)$$

which signifies the negative power of 10 for $[H^+]$, eqn (1.50) becomes

$$p_{A^-} = \frac{1}{1 + 10^{pK-pH}},$$

$$p_{AH} = \frac{10^{pK-pH}}{1 + 10^{pK-pH}} = \frac{1}{10^{pH-pK} + 1}. \quad (1.68)$$

SIMPLE MULTIPLE EQUILIBRIA

For pH = $pK \pm 1$ and pH = pK we obtain

	p_{A^-}	p_{AH}
pH = $pK - 1$	$\frac{1}{1+10} \sim 0.1$	$\frac{1}{1+\frac{1}{10}} \sim 0.9$
pH = pK	$\frac{1}{2}$	$\frac{1}{2}$
pH = $pK + 1$	$\frac{1}{1+\frac{1}{10}} \sim 0.9$	$\frac{1}{1+10} \sim 0.1$

From the above tabulation we see that the transition from approximately all AH to all A^- takes place over two pH units centred about the value pH = pK. It is important to realize that eqn (1.68) and the above tabulation are independent of the particular value of pK and that the behaviour of all monobasic acids as a function of pH is identical, the curves simply being shifted so as to be centred about the proper pK. Table 1.3 lists the pKs of some of the important monobasic acids,[5] and Fig. 1.6 shows how the probabilities of representative species vary as a function of pH. Notice that in the pH region pH = 7 ± 1 amino and carboxyl groups are going to be in the charged forms RNH_3^+ and $RCOO^-$ respectively.

As an example of a dibasic acid consider carbonic acid. In any introductory chemistry text we will find the following two reactions listed,

$$H_2CO_3 \rightleftharpoons HCO_3^- + H^+, \quad K_{d1} = 10^{-pK_1} = 10^{-6.37},$$
$$HCO_3^- \rightleftharpoons CO_3^{2-} + H^+, \quad K_{d2} = 10^{-pK_2} = 10^{-10.25}, \quad (1.69)$$

TABLE 1.3

Proton dissociation constants for some monobasic acids at 298 K

Acid	pK	Acid	pK
formic	3.75	(NH_4^+ Cl^-)	9.25
acetic	4.75	methylamine(H^+)	10.64
propionic	4.87	ethylamine(H^+)	10.75
n-butyric	4.81	n-butylamine(H^+)	10.61
lactic	3.08		
acetoacetic	3.58		

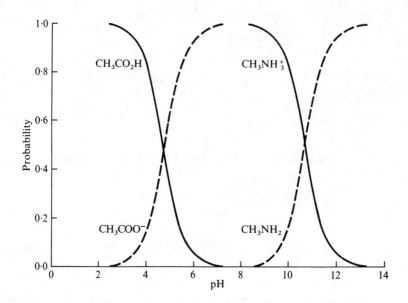

FIG. 1.6. Probability distribution for the states of proton binding of methylamine and acetic acid as a function of pH.

which we immediately rewrite as

$$CO_3^{2-} + H^+ \rightleftharpoons HCO_3^-, \quad K_1 = 1/K_{d1} = 10^{pK_1^*} = 10^{6.37},$$
$$CO_3^{2-} + 2H^+ \rightleftharpoons H_2CO_3, \quad K_2 = 1/K_{d1}K_{d2} = 10^{pK_2^*} = 10^{16.62}. \quad (1.70)$$

Then

$$p_{CO_3} = 1/\xi,$$
$$p_{HCO_3^-} = \left(10^{pK_1^* - pH}\right)/\xi,$$
$$p_{H_2CO_3} = \left(10^{pK_2^* - 2pH}\right)/\xi, \quad (1.71)$$
$$\xi = 1 + 10^{pK_1^* - pH} + 10^{pK_2^* - 2pH}$$

The symbol pK^* indicates our convention of writing associations (eqn (1.70)) in contrast to the conventional form of writing dissociations (eqn (1.69)). For this example the two pKs are

SIMPLE MULTIPLE EQUILIBRIA

separated by 4 pH units. Since we have seen for a monobasic acid that the transition from AH to A^- takes place almost completely over 2 pH units, the results of eqn (1.71) are accurately given by treating eqn (1.69) as two independent processes. That is, if the pKs are separated by more than 2 pK units, then to a very good approximation there are at most two species of any significant probability present at any pH. Fig. 1.7 shows the probabilities listed in eqn (1.71) as a function of pH; we see that the results are well separated curves of the same form as those given for monobasic acids in Fig. 1.6.

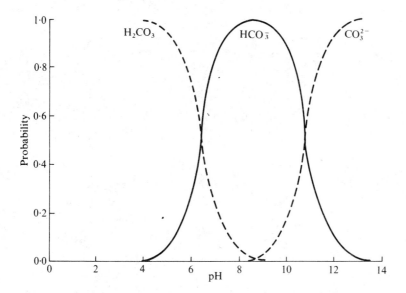

FIG. 1.7. Probability distribution for the states of proton binding of carbonic acid as a function of pH.

Phosphoric acid has three protons that can dissociate and is characterized by three stepwise reactions

$$H_3PO_4 \rightleftharpoons H_2PO_4^- + H^+, \quad K_{d1} = 10^{-pK_1} = 10^{-2.12},$$

$$H_2PO_4^- \rightleftharpoons HPO_4^{2-} + H^+, \quad K_{d2} = 10^{-pK_2} = 10^{-7.21}, \quad (1.72)$$

$$HPO_4^{2-} \rightleftharpoons PO_4^{3-} + H^+, \quad K_{d3} = 10^{-pK_3} = 10^{-12.67},$$

or writing all the species relative to PO_4^{-3}

$$PO_4^{3-} + H^+ \rightleftharpoons HPO_4^{2-}, \quad K_1 = 1/K_{d3} = 10^{pK_1^*} = 10^{12.67}.$$

$$PO_4^{3-} + 2H^+ \rightleftharpoons H_2PO_4^-, \quad K_2 = 1/K_{d2}K_{d3} = 10^{pK_2^*} = 10^{19.88}. \quad (1.73)$$

$$PO_4^{3-} + 3H^+ \rightleftharpoons H_3PO_4, \quad K_3 = 1/K_{d1}K_{d2}K_{d3} = 10^{pK_3^*} = 10^{22.00}.$$

The probability distribution is then given by

$$p(PO_4^{3-}) = 1/\xi,$$

$$p(HPO_4^{3-}) = \left(10^{pK_1^* - pH}\right)/\xi,$$

$$p(H_2PO_4^-) = \left(10^{pK_2^* - 2pH}\right)/\xi, \quad (1.74)$$

$$p(H_3PO_4) = \left(10^{pK_3^* - 3pH}\right)/\xi,$$

$$\xi = 1 + 10^{pK_1^* - pH} + 10^{pK_2^* - 2pH} + 10^{pK_3^* - 3pH}.$$

Fig. 1.8 shows the probability of each species of phosphoric acid as a function of pH; again for this example the pKs of the stepwise dissociation (eqn (1.71)) are separated by more than two pH units, and hence for phosphoric acid there are at most two species of any significant probability present at any pH and the net distribution is accurately given by the superposition of the steps given in eqn (1.72) treated independently.

To illustrate how simple eqn (1.74) makes the treatment of proton dissociation, consider the terms in ξ at pH = 7:

$$\xi = 1 + 10^{12.67-7} + 10^{19.88-14} + 10^{22.00-21}$$

$$= 1 + 10^{5.67} + 10^{5.88} + 10^{1.00}$$

$$= \begin{pmatrix} \text{term for no} \\ \text{protons} \\ \text{bound} \end{pmatrix} + \begin{pmatrix} \text{term for one} \\ \text{proton} \\ \text{bound} \end{pmatrix} + \begin{pmatrix} \text{term for two} \\ \text{protons} \\ \text{bound} \end{pmatrix} + \begin{pmatrix} \text{term for three} \\ \text{protons} \\ \text{bound} \end{pmatrix}.$$

The terms in ξ are directly proportional to the probability of the various states of binding. In the above example it is

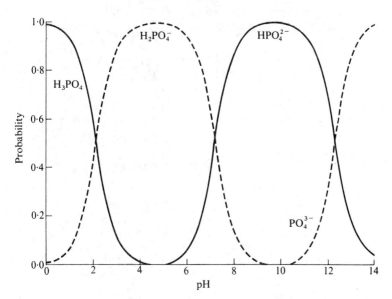

FIG. 1.8. Probability distribution for the states of proton binding of phosphoric acid as a function of pH.

immediately seen that only the species HPO_4^{2-} and $H_2PO_4^-$ are significantly probable at pH = 7.

From eqns (1.68), (1.71), and (1.74), the reader will see that from a practical viewpoint all we need do is write all the equilibria relative to the species with no protons bound

$$A_0 + nH^+ \rightleftharpoons A_n, \qquad K_n = 10^{pK_n^*}. \tag{1.75}$$

Then immediately

$$p_{A_n} = \left(10^{pK_n^* - npH}\right) \bigg/ \sum_{n=0}^{n} 10^{pK_n^*} - npH, \tag{1.76}$$

which simply involves the evaluation of powers of 10. The only complication in using this approach is that the reactions must be written as in eqn (1.75), which is not the form generally found in books; as shown, for example, in eqn (1.73) it is a simple matter to convert from conventional pKs to the pK^* used here.

We notice further that the probability distribution of eqn (1.76) has the same general form as that given for the probability distribution for simple quantum-mechanical energy levels (see eqn (1.13)), namely

$$p(\text{species}) = \frac{\text{term(species)}}{\Sigma \text{ term(species)}} . \qquad (1.77)$$

As an example of a tribasic acid where the pKs are very close together consider citric acid, which has three carboxyl groups. The equilibria are

$$\begin{aligned}
CtH_3 &\rightleftharpoons CtH_2^- + H^+, & K_{d1} &= 10^{-3.08}, \\
CtH_2^- &\rightleftharpoons CtH^{2-} + H^+, & K_{d2} &= 10^{-4.74}, \\
CtH^{2-} &\rightleftharpoons Ct^{3-} + H^+, & K_{d3} &= 10^{-5.4}, \\
Ct^{3-} + H^+ &\rightleftharpoons CtH^{2-}, & K_1 &= 10^{5.4}, \\
Ct^{3-} + 2H^+ &\rightleftharpoons CtH_2^-, & K_2 &= 10^{10.14}, \\
Ct^{3-} + 3H^+ &\rightleftharpoons CtH_3, & K_3 &= 10^{13.22},
\end{aligned} \qquad (1.78)$$

Fig. 1.9 gives the species probability distribution for this molecule as a function of pH (the equations involved are eqn (1.74) with the pK^* of eqn (1.78)). We see that when the pKs are closer than 2 pH units the species probabilities considerably overlap and we cannot estimate the probabilities accurately by treating each dissociation step independently.

Now consider a polybasic acid, for example, with N carboxyl groups where the dissociation of each carboxyl group is independent of the states of association of the others with the same intrinsic dissociation constant for each. The quantity ξ is given formally by

$$\xi = \sum_{n=0}^{N} [H^+]^n K_n, \qquad (1.79)$$

but since we are assuming that the carboxyl groups are independent and equivalent, it is clear that all the K_n can be

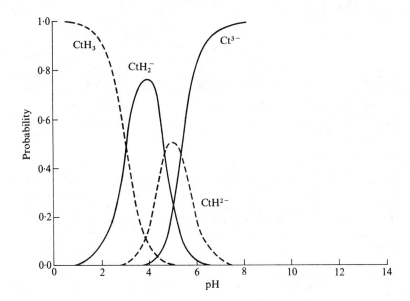

FIG. 1.9. Probability distribution for the states of proton binding of citric acid as a function of pH.

written in terms of the single association constant K characterizing each group. If we write the association reaction for 4 specific carboxyl groups, then

$$\begin{array}{c} \text{COO}^- \\ | \\ \text{COO}^- \\ | \\ \text{COO}^- \\ | \\ \text{COO}^- \end{array} + 4\text{H}^+ \rightleftharpoons \begin{array}{c} \text{CO}_2\text{H} \\ | \\ \text{CO}_2\text{H} \\ | \\ \text{CO}_2\text{H} \\ | \\ \text{CO}_2\text{H} \end{array} \quad K^4. \quad (1.80)$$

Suppose that there are a total of 16 carboxyl groups on the molecule; then there are a large number of ways to pick four specific groups to form a reaction like that in eqn (1.80), all of which have equilibrium constant K^4. In fact there are

$$\frac{16(16-1)(16-2)(16-3)}{(4)(3)(2)(1)} = \frac{16!}{4!(16-4)!}$$

distinct ways we can put 4 protons on 16 groups. The generalization is that there are

$$\frac{N!}{n!(N-n)!}$$

ways to place n protons on N carboxyl groups each reaction having equilibrium constant K^n. Then the net equilibrium constant for the sum of all the ways to place n protons on N carboxyl groups is

$$N\text{COO}^- + n\text{H}^+ \rightleftharpoons (N-n)\text{COO}^- + n\text{CO}_2\text{H} \tag{1.81}$$

$$K_n = \frac{N!}{n!(N-n)!} K^n.$$

Then ξ in eqn (1.79) becomes

$$\xi = \sum_{n=0}^{N} \frac{N!}{n!(N-n)!} [\text{H}^+]^n K^n. \tag{1.82}$$

Using the formula for the binomial expansion

$$(x+y)^N = \sum_{n=0}^{N} \frac{N!}{n!(N-n)!} x^n y^{N-n}, \tag{1.83}$$

we obtain a closed form for ξ

$$\xi = (1+[\text{H}^+]K)^N. \tag{1.84}$$

Eqn (1.84) can be interpreted in terms of the fact that there are two states for each carboxyl group, COO^- represented by the factor 1 and CO_2H by the factor $[\text{H}^+]K$; all possible states of N such independent groups are generated by the product $(1+[\text{H}^+]K)^N$.

To calculate the average number of protons bound to the polybasic acid we use eqn (1.64) and obtain

$$\langle n \rangle = N \frac{[\text{H}^+]K}{1 + [\text{H}^+]K}. \tag{1.85}$$

We can introduce the notion of the fraction ϑ of carboxyl groups with bound protons by dividing $\langle n \rangle$ by the maximum number N of protons that can possibly bind, giving

SIMPLE MULTIPLE EQUILIBRIA 31

$$\vartheta = \frac{\langle n \rangle}{N} = \frac{[H^+]K}{1 + [H^+]K}, \qquad (1.86)$$

which is identical with the relation we derived for the case of a monobasic acid (eqn (1.50) or eqn (1.68)). This is, of course, physically what one would expect since we assumed the carboxyl groups were independent and equivalent, hence the average state of binding of all of them is given by examining a single representative group or, in other words, the fact that the groups are all tied to the same molecule makes no difference, they bind protons as if each were free in solution. Of course, we assumed they were independent and equivalent, and the above results simply point up the consistency of our mathematical treatment.

With the above approach in mind let us re-examine dibasic acids using the dicarboxylic acids, HOOCXYCOOH, as an example, where the symbol XY indicates that the environment of the two carboxyl groups is different. The equilibria are

$$\begin{aligned}
^-OOC-XY-COO^- + H^+ &\rightleftharpoons HOOC-XY-COO^-, & K_{1a}, \\
^-OOC-XY-COO^- + H^+ &\rightleftharpoons {}^-OOC-XY-COOH, & K_{1b}, \\
^-OOC-XY-COO^- + 2H^+ &\rightleftharpoons HOOC-XY-COOH, & K_2.
\end{aligned} \qquad (1.87)$$

In terms of the reactions given in eqn (1.87), ξ is given as

$$\begin{aligned}
\xi &= 1 + [H^+]K_{1a} + [H^+]K_{1b} + [H^+]^2 K_2 \\
&= 1 + [H^+](K_{1a} + K_{1b}) + [H^+]^2 K_2.
\end{aligned} \qquad (1.88)$$

Comparing eqn (1.88) with the last of eqn (1.71), which we rewrite here as

$$\xi = 1 + [H^+]K_1 + [H^+]^2 K_2, \qquad (1.89)$$

we find

$$K_1 = K_{1a} + K_{1b}. \qquad (1.90)$$

The constants K_1 and K_2 can be determined experimentally from matching the theoretical titration curve (eqn (1.64)) using the ξ of eqn (1.89)) with the experimental one. Since we can only determine K_1 experimentally, it is impossible with no special assumptions to determine K_{1a} and K_{1b} and, hence, the probability of the two species HOOC–XY–COO$^-$ and $^-$OOC–XY–COOH separately (the combined probability can, of course, be determined). If the dicarboxylic acid is symmetric (X=Y), which is the case for HOOC(CH$_2$)$_n$COOH, then $K_{1a} = K_{1b}$ and $K_1 = 2K_{1a}$. For this special case, if n becomes large then we would expect the dissociation of each carboxyl group to become independent of the state of ionization of the other. In that case HOOC(CH$_2$)$_n$COOH becomes a special case ($N=2$) of the polybasic acid containing N independent carboxyl groups treated previously. Eqn (1.84) with $N = 2$ gives

$$\xi = (1+[H^+]K)^2 = 1 + 2[H^+]K + [H^+]^2 K^2. \quad (1.91)$$

Comparing eqns (1.91) and (1.88) we find

$$K_{1a} = K = K_{1b}, \qquad K_2 = K^2. \quad (1.92)$$

Table 1.4 gives pK_1 and pK_2 for the stepwise dissociation (as in eqn (1.69)) and the constants pK_{1a}^* and pK_2^* (the positive powers of 10 for our K_{1a} and K_2 respectively) for the symmetric

TABLE 1.4.

Proton dissociation constants for the symmetric dicarboxylic acids HOOC(CH$_2$)$_n$COOH at 298 K

n	Name	pK_{d1}	pK_{d2}	pK_{1a}^*	pK_2^*
0	oxalic	1.23	4.19	0.93	5.42
1	malonic	2.83	5.69	2.53	8.52
2	succinic	4.16	5.61	3.86	9.77
3	glutaric	4.34	5.41	4.04	9.75
4	adipic	4.41			
	acetic	4.75			

SIMPLE MULTIPLE EQUILIBRIA   33

carboxylic acids for n from 1 to 4. From eqn (1.92), if the two carboxyl groups are independent then pK_1^* and $\frac{1}{2}pK_2^*$ should be the same and be equal to the pK for an independent carboxyl group, e.g., that of acetic acid. We see in Table 1.4 that as n becomes large the two carboxyl groups do indeed become independent. The main reason for the fact that the carboxyl groups are not independent for small n is the electrostatic interaction between the carboxyl groups at small distances.

All of the equations we have given for treating multiple binding of protons apply equally well to the binding of other ions and small molecules. For example, if B is any species that can bind to A then

$$A + B \rightleftharpoons A\text{-}B, \qquad K,$$

$$p_A = \frac{1}{1 + [B]K}, \qquad (1.94)$$

$$p_B = \frac{[B]K}{1 + [B]K},$$

which are, of course, exactly the same as eqns (1.44) and (1.50).

1.8. Equilibria between free energy levels

So far we have discussed equilibrium in three types of system: equilibria between quantum-mechanical energy levels; equilibria between free energy levels for different conformational isomers of a molecule; and equilibria between states of binding for polybasic acids. Summarizing the important equations

quantum-mechanical energy levels	$p_n = \exp(-\varepsilon_n/RT)/\Sigma \exp(-\varepsilon_n/RT),$	(1.95a)
free energy levels	$p_n = \exp(-g_n^0/RT)/\Sigma \exp(-g_n^0/RT),$	(1.95b)
states of binding	$p_n = [H^+]^n K_n / \Sigma [H^+]^n K_n,$	(1.95c)

we see that the mathematical form of the distributions is the same (the general form of which is given by eqn (1.77)). We now inquire into the relation between the parameters which occur in eqn (1.95).

First we note that the quantity $[H^+]^n K_n$ can be written as

$$[H^+]^n K_n = [H^+]^n \exp(-g_n^0/RT) = \exp\{-(g_n^0 + nRT \ln[H^+])/RT\}, \tag{1.96}$$

where g_n^0 is the standard free energy for the state with n protons bound relative to n dissociated protons and the completely dissociated molecule. Thus taking

$$g_n = g_n^0 + nRT \ln[H^+], \tag{1.97}$$

eqn (1.95c) becomes

$$p_n = \exp(-g_n/RT)/\Sigma \exp(-g_n/RT), \tag{1.98}$$

which is exactly the same form as eqn (1.95b) for equilibrium between free energy levels g_n^0. But through eqn (1.97) the g_n of eqn (1.98) depend on the hydrogen-ion concentration. Thus multiple binding of protons (or any other species) can be viewed as a multiple equilibria between free energy levels the position and spacing of which are functions of the concentration of the binding species.

As was mentioned earlier, the fundamental difference between eqn (1.95a) and (1.95b) is that the ε_n in eqn (1.95a) are constants for a given system (for fixed volume) while the g_n^0 of eqn (1.95b) are functions of temperature. By definition

$$g_n^0 = h_n^0 - T s_n^0, \tag{1.99}$$

where h_n^0 and s_n^0 are respectively the standard enthalpy and entropy per mole. While in principle the enthalpy and entropy are functions of temperature, over a temperature range of 300–310 K it is commonly a very good approximation to treat the enthalpy and entropy as constants. With this approximation the temperature dependence of the factors given in eqn (1.95b) is then given explicitly as

$$\exp(-g_n^0/RT) = \exp(s_n^0/R) \exp(-h_n^0/RT), \tag{1.100}$$

which has exactly the same temperature dependence as the corresponding factor in a degenerate energy-level scheme, namely $\omega \exp(-\varepsilon_n/RT)$, with the identification (see eqn (1.18))

$$h_n^0 \to \varepsilon_n,$$
$$s_n^0 \to R\ln\omega . \qquad (1.101)$$

The quantity $\exp(-g_n/RT)$ in eqn (1.98) using the same approximation of constant enthalpy and entropy has the temperature dependence

$$\exp(-g_n/RT) = [H^+]^n \exp(s_n^0/R) \exp(-h_n^0/RT), \qquad (1.102)$$

which is also analogous to the case of a degenerate-energy-level scheme with the identification

$$h_n^0 \to \varepsilon_n,$$
$$s_n^0 + R\ln[H^+] \to R\ln\omega. \qquad (1.103)$$

Thus, over a limited temperature range, where we can assume that the enthalpy and entropy are constant, equilibria between the free energy levels g_n^0 or g_n can be thought of as the exact analogue of equilibrium between degenerate energy levels.

In the next chapter we will pursue further the relation between free energy levels and quantum-mechanical energy levels.

1.9. Physical cluster model

We have examined general relations for multiple equilibria between conformational isomers and different states of association of polybasic acids. We now turn to a general treatment of multiple self-association, where molecules can associate with one another to form dimers, trimers, etc.[6] We follow our previous convention in writing all equilibria with respect to the completely dissociated state. We have

$$A_1 \rightleftharpoons A_1, \quad K_1 = 1$$
$$2A_1 \rightleftharpoons A_2, \quad K_2$$
$$3A_1 \rightleftharpoons A_3, \quad K_3 \tag{1.104}$$
$$\cdots$$
$$nA_1 \rightleftharpoons A_n, \quad K_n$$
$$\cdots$$

The law of mass action applied to the general term of eqn (1.104) gives

$$\frac{[A_n]}{[A_1]^n} = K_n \tag{1.105}$$

or

$$[A_n] = [A_1]^n K_n. \tag{1.106}$$

The conservation of monomer units gives the relation

$$\sum_{n=1}^{\infty} n[A_n] = c_0. \tag{1.107}$$

Using eqn (1.106) for A_n, eqn (1.107) becomes

$$\sum_{n=1}^{\infty} n[A_1]^n K_n = c_0, \tag{1.108}$$

which if the K_n are known is an equation for the single unknown $[A_1]$. Solving eqn (1.108) for $[A_1]$ (which must be done numerically for arbitrary K_n), all the $[A_n]$ are then given by eqn (1.106) in terms of $[A_1]$ and the K_n.

The total concentration of clusters of any size is given by

$$C = \sum_{n=1}^{\infty} [A_n] = \sum_{n=1}^{\infty} [A_1]^n K_n. \tag{1.109}$$

For dilute solutions, C can be measured by using any physical parameter, such as osmotic pressure or the freezing-point depression, that is dependent only on the number of particles (clusters). The probability distribution of cluster size is then given by

SIMPLE MULTIPLE EQUILIBRIA

$$p_n = [A_n]/C = \frac{[A_1]^n K_n}{\Sigma [A_1]^n K_n}, \qquad (1.110)$$

which is analogous to eqn (1.62), except that in general $[A_1]$, the concentration of monomer, is not experimentally available, this quantity being given by the solution of eqn (1.108) in terms of the known quantity c_0 (the total concentration of monomer units in any form). Eqns (1.108) to (1.110) have some extraordinary properties which are not immediately evident. To explore these properties we will examine some special forms of the K_n.

The first special case we examine is for linear association having the same association constant for each step of association

$$\begin{aligned} A_1 + A_1 &\rightleftharpoons A_2, & K, \\ A_2 + A_1 &\rightleftharpoons A_3, & K, \\ A_3 + A_1 &\rightleftharpoons A_4, & K, \\ &\cdots \\ A_{n-1} + A_1 &\rightleftharpoons A_n, & K, \\ &\cdots \end{aligned} \qquad (1.111)$$

By adding appropriate reactions from eqn (1.111) to form the reactions of eqn (1.104) we find the relation

$$K_n = K^{n-1}. \qquad (1.112)$$

For the special form of the K_n given in eqn (1.112), eqn (1.108) becomes

$$\sum_{n=1}^{\infty} n x^n = K c_0, \qquad (1.113)$$

with

$$x = [A_1] K. \qquad (1.114)$$

We have already met the sum appearing in eqn (1.113) in eqn (1.23). Thus

$$x/(1-x)^2 = Kc_0. \tag{1.115}$$

Since eqn (1.115) is quadratic in x, we can obtain an explicit solution for x (and hence $[A_1]$). However, a graphical solution yields more insight into the nature of this multiple equilibria.

Fig. 1.10 shows $x/(1-x)^2$ as a function of x; the intersection of this curve with the constant Kc_0 gives the solution.

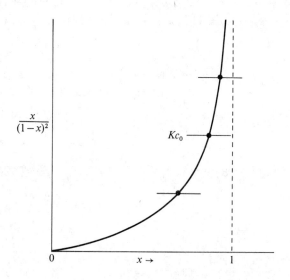

FIG. 1.10. Schematic illustration of the graphical solution of eqn (1.115). The short horizontal lines indicate various values of Kc_0, the dot at the intersection of the two curves being the desired solution. Note that $x = 1$ is not an allowed solution for finite Kc_0 and that x must be less than unity.

Since $x/(1-x)^2$ goes to infinity as x approaches unity, we see immediately that

$$x < 1,$$
$$[A_1] < K^{-1}, \tag{1.116}$$

no matter what the value of c_0. This means that $[A_1]$ has the upper bound K^{-1} that cannot be exceeded no matter how concen-

SIMPLE MULTIPLE EQUILIBRIA

trated the solution. Of course, this specific conclusion depends on the form of K_n given in eqn (1.112). The probability distribution (eqn (1.110)), has the simple form

$$p_n = x^n / \sum_{n=1}^{\infty} x^n. \qquad (1.117)$$

The sum in eqn (1.117) is again the geometric series

$$\sum_{n=1}^{\infty} x^n = x/(1-x)$$

(compare eqn (1.115) which is a sum from zero to infinity) giving, for eqn (1.117),

$$p_n = (1-x)x^{n-1} \quad (n=1,\infty), \qquad (1.118)$$

which has exactly the same form as the probability distribution for the multilevel energy scheme with equal spacing (eqn (1.16), where $n=0,\infty$). Hence the nature of the distribution as a function of x (from eqn (1.116), $x < 1$) has already been given in Fig. 1.3. Since $x < 1$, p_n is a monotonically decreasing function of x. Hence, for all c_0, the monomer is the most probable cluster size.

We must keep in mind that eqn (1.118) gives the probability distribution for cluster sizes. If we want the distribution for the probability a molecule is in a cluster of n molecules then this probability, which we designate P_n, is proportional to the probability of having a cluster of a given size times the number of particles in a cluster

$$P_n \sim n p_n. \qquad (1.119)$$

Using eqn (1.118) for p_n, we obtain

$$P_n = (1-x)^2 n x^{n-1}, \qquad (1.120)$$

where we have evaluated the proportionality constant in eqn (1.119) so as to give

$$\sum_{n=1}^{\infty} P_n = 1, \qquad (1.121)$$

which we may check using eqn (1.23). This distribution is analogous to the degenerate multilevel system treated earlier (see eqn (1.24)) and, hence, the P_n go through a maximum for an intermediate value of n; the behaviour of P_n as a function of x and n is identical to that shown in Fig. 1.5. It is of interest to calculate the value of n for which P_n is a maximum. We may estimate this by treating n as a continuous variable and set $\partial P_n/\partial n = 0$ (or $\partial\{n\exp(n\ln x)\}/\partial n = 0$). Doing this we obtain

$$n_{max} = 1/\ln x^{-1}. \tag{1.122}$$

To obtain a more useful estimate of n_{max} we first evaluate the total concentration C of clusters of eqn (1.109)

$$C = K^{-1} \sum_{n=1}^{\infty} x^n = K^{-1} x/(1-x). \tag{1.123}$$

Then the ratio of C to c_0 is given by (using eqns (1.115) and (1.123))

$$C/c_0 = 1 - x \tag{1.124}$$

or

$$x = 1 - C/c_0. \tag{1.125}$$

If the extent of clustering is large, then $C/c_0 \ll 1$ and

$$\ln x^{-1} = -\ln x = -\ln(1 - C/c_0) \simeq C/c_0,$$

giving for n_{max} of eqn (1.122)

$$n_{max} \simeq c_0/C. \tag{1.126}$$

Note that c_0/C (the total monomer concentration in any form divided by the total cluster concentration) is simply the average number of monomer units in a cluster. In any association or polymerization scheme it is very important to keep the two distributions p_n (probability of cluster size) and P_n (probability a monomer unit is in a cluster of a given size)

SIMPLE MULTIPLE EQUILIBRIA

distinct: they have quite different behaviours (see Figs 1.3 and 1.5 respectively).

While we are discussing linear association, it is useful to introduce a modification of the equilibrium scheme given in eqn (1.111). In the scheme of eqn (1.111) any two monomer units can join and start a cluster. The modification we consider now is where the number of clusters is fixed, that is association can only take place on initiator molecules I. The equilibrium scheme is

$$
\begin{aligned}
I + A &\rightleftharpoons IA_1, \quad K, \\
IA_1 + A &\rightleftharpoons IA_2, \quad K, \\
&\cdots \\
IA_{n-1} + A &\rightleftharpoons IA_n, \quad K, \\
&\cdots
\end{aligned}
\tag{1.127}
$$

or

$$I + nA \quad IA_n, \quad K_n = K^n. \tag{1.128}$$

Applying the law of mass action we have

$$[IA_n] = [I][A]^n K^n, \tag{1.129}$$

which is analogous to eqn (1.106), except that the cluster concentration $[IA_n]$ is expressed in terms of two variables $[I]$ and $[A]$, these being respectively the concentrations of unreacted initiator and unreacted monomer. The conservation of total monomer and initiator units gives us two equations in two unknowns (analogues of eqns (1.108) and (1.109) respectively)

conservation of monomer units
$$[A] + \sum_{n=1}^{\infty} n[IA_n] = c_0, \tag{1.130}$$

conservation of initiator
$$\sum_{n=0}^{\infty} [IA_n] = I_0. \tag{1.131}$$

Defining

$$\begin{aligned} x &= [A]K, \\ y &= [I], \end{aligned} \tag{1.132}$$

then, with eqns (1.132) and (1.129), eqns (1.130) and (1.131) become

$$K^{-1}x + y \sum_{n=1}^{\infty} nx^n = c_0$$
$$y \sum_{n=0}^{\infty} x^n = I_0, \qquad (1.133)$$

or using the geometric series

$$K^{-1}x + yx/(1-x)^2 = c_0,$$
$$y/(1-x) = I_0. \qquad (1.134)$$

Eliminating y we obtain an equation just in terms of x,

$$x/I_0 K + x/(1-x) = c_0/I_0. \qquad (1.135)$$

It is clear from eqn (1.135) that x must be less than unity (this is also immediately evident from eqns (1.133) which diverge for $x \geqslant 1$). Hence we have an upper bound on [A]

$$x < 1,$$
$$[A] < K^{-1}. \qquad (1.136)$$

Eqn (1.135) is quadratic in x and hence is readily solved explicitly for x in terms of $I_0 K$ and c_0/I_0. The cluster distribution function for this model is

$$p_n = [IA_n]/I_0 = (1-x)x^n \quad (n=0,\infty), \qquad (1.137)$$

which, since x is less than unity, is always a monotonically decreasing function of x (exactly as in Fig. 1.3) no matter what the values of K, c_0, and I_0. Note that the concentration of unreacted initiator is simply $(1-x)I_0$.

The model we have been discussing of chain growth on initiator molecules is applicable to the *in vitro* enzymatic synthesis of certain synthetic polynucleotides[7] and starches.[8] Apparently the enzymes can only catalyse the addition of monomer efficiently when there is a nucleus of polymer to which it can bind.

SIMPLE MULTIPLE EQUILIBRIA   43

The general model of self-associating molecules contained in eqns (1.104) to (1.110) is the general starting point for treating many aggregating systems of interest in biophysical chemistry. Having examined the specific properties of eqns (1.104) to (1.110) for linear association to form chains, we now examine the appropriate forms the K_n take for such specifically shaped aggregates as spherical clusters and round platelets, which are of interest with regard to micelles and films (membrane models). It is not our place here to discuss the fine details of particular systems but rather to indicate how the multiple equilibria aspect of the problem can be handled most simply.

For the model of linear association just treated, the general K_n of eqn (1.104) were all expressed in terms of the single constant K for addition of a monomer unit to a chain (eqns (1.112) and (1.128)), thus reducing the number of equilibrium parameters to one. For other shape aggregates we want to follow the same idea and describe all the K_n in terms of just a few parameters. Let us first consider spherical aggregates. We have the simple geometric properties

$$\begin{aligned}
V_n &= \text{volume of aggregate of } n \text{ particles} = nv_0, \\
v_0 &= \text{volume occupied in an aggregate by a single particle}, \\
V_n &= \tfrac{4}{3}\pi r^3, \\
r &= \text{radius of aggregate}, \\
S_n &= \text{surface area of aggregate} = 4\pi r^2
\end{aligned} \tag{1.138}$$

Using the above properties we have

$$\begin{aligned}
V_n &\sim n, \\
r &\sim n^{\frac{1}{3}}, \\
S_n &\sim n^{\frac{2}{3}}.
\end{aligned} \tag{1.139}$$

If we make the approximation that there are two distinctly different environments of a molecule in a spherical aggregate, namely the interior, where the molecule is completely surrounded by other molecules like itself, and the surface, where a molecule is partly surrounded by other like molecules in the cluster and partly surrounded by solvent, then there are two free energies that characterize a spherical aggregate,

$$\Delta g_i^0 = g_i^0 \begin{pmatrix} \text{molecule in} \\ \text{interior} \end{pmatrix} - g^0 \begin{pmatrix} \text{molecule free} \\ \text{in solution} \end{pmatrix},$$
$$\Delta g_s^0 = g_s^0 \begin{pmatrix} \text{molecule on} \\ \text{the surface} \end{pmatrix} - g_i^0 \begin{pmatrix} \text{molecule in} \\ \text{interior} \end{pmatrix}. \qquad (1.140)$$

Then using eqns (1.140) and (1.139) we have

$$K_n = \exp(-an^{\frac{2}{3}} \Delta g_s^0 / RT) \exp(-bn \, \Delta g_i^0 / RT), \qquad (1.141)$$

where a and b are proportionality constants that make eqns (1.139) equalities (these being simple geometric factors). For the K_n of eqn (1.141), eqn (1.108) has the form

$$\sum_{n=1}^{\infty} n [A_1]^n \exp(-an^{\frac{2}{3}} \Delta g_s^0 / RT) \exp(-bn \, \Delta g_i^0 / RT) = c_0. \qquad (1.142)$$

To emphasize the n dependence of eqn (1.142) we define

$$x = [A_1] \exp(-b \, \Delta g_i^0 / RT),$$
$$y = \exp(-a \, \Delta g_s^0 / RT). \qquad (1.143)$$

Using the definitions of eqn (1.143), eqn (1.142) is simply

$$\sum_{n=1}^{\infty} n y^{n^{\frac{2}{3}}} x^n = c_0. \qquad (1.144)$$

Owing to the $n^{\frac{2}{3}}$ power of y eqn (1.144) is no longer the geometric series, and we cannot obtain a simple closed form for the sum. If we know Δg_i^0 and Δg_s^0, eqn (1.44) contains only one unknown quantity $[A_1]$, and the solution can be obtained numerically by evaluating the sum on a computer at different values of x, plotting the sum as a function of x, the solution being

SIMPLE MULTIPLE EQUILIBRIA

the intersection with the constant c_0. However, the qualitative features of eqn (1.144) can be deduced without having a closed form for the sum. And these features are remarkable for such a simple model.

First, we note the convergence properties of the sum:

$$
\begin{aligned}
&x > 1, \quad y > 0 \quad &&\text{sum diverges} \\
&x = 1, \quad y \geqslant 1 \quad &&\text{sum diverges} \\
&x = 1, \quad y < 1 \quad &&\text{sum converges} \\
&x < 1, \quad y < 1 \quad &&\text{sum converges} \\
&x < 1, \quad y \geqslant 1 \quad &&\text{sum converges.}
\end{aligned}
\qquad (1.145)
$$

From expressions (1.145) we immediately conclude

$$
\begin{aligned}
&x \leqslant 1 \quad \text{if } y < 1, \\
&x < 1 \quad \text{if } y \geqslant 1.
\end{aligned}
\qquad (1.146)
$$

Thus x must always be less than or equal to unity, where $x = 1$ is a possible solution if $y < 1$. To see what $x = 1$ implies physically, let us investigate the meaning of $y < 1$. If physical clusters are favoured (i.e. have a tendency to form) then we expect that $\Delta g_i^0 < 0$ (recall that a negative free-energy change favours the reaction as written here (eqn (1.140a)) from solution to the interior). A molecule on the surface is only partly surrounded by like molecules and hence is in a high-free-energy situation; it has lost the translational entropy of being free in solution, but has only partly gained interactions with other molecules of its kind to compensate for this loss. The following three conditions are concomitant:

$$
\left.\begin{aligned}
&\text{surface unfavourable,} \\
&\Delta g_s^0 > 0, \\
&y < 1.
\end{aligned}\right\}
\qquad (1.147)
$$

Incidently, it is necessary for Δg_s^0 to be greater than zero (unfavourable surface free energy) in order for the assumption of spherical aggregates to make sense since the spherical shape is the geometric form that minimizes the amount of surface. If the surface was favourable then aggregation would produce an all-surface type of structure resulting in a sponge-like labyrinth for which it would be very difficult to formulate a simple recipe for the K_n.

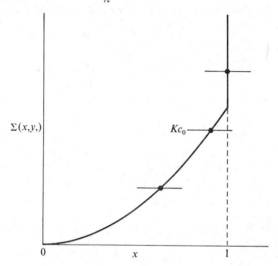

FIG. 1.11. Schematic illustration of the graphical solution of eqn (1.144). As with Fig. 1.10, the short horizontal lines indicate various values of Kc_0. In contrast to the behaviour shown in Fig. 1.10, $x = 1$ is an allowed solution for finite Kc_0, although we still have $x \leq 1$.

Fig. 1.11 shows the sum of eqn (1.144) plotted schematically as a function of x for $y < 1$ with various values of c_0 indicated. Since the solution of eqn (1.144) is the intersection of the sum plotted as a function of x with c_0, we see immediately that there is a particular value c_0^* of c_0 above which $x = 1$. Thus

$$x = 1 \text{ for } c_0 \geq c_0^*. \tag{1.148}$$

Thus, for the conditions of expression (1.148), we have a solution of eqn (1.144). For $x = 1$ the cluster distribution, eqn (1.110) is given by

SIMPLE MULTIPLE EQUILIBRIA

$$p_n = \left(y^{n^{\frac{2}{3}}}\right) \bigg/ \sum_{n=1}^{\infty} y^{n^{\frac{2}{3}}}. \tag{1.149}$$

Since $y < 1$, p_n is a monotonically decreasing function of n. The total concentration of clusters at $x = 1$ (eqn (1.109)) is

$$C = \sum_{n=1}^{\infty} y^{n^{\frac{2}{3}}}. \tag{1.150}$$

Eqns (1.149) and (1.150) are extraordinary results for they mean that for any total monomer concentration $c_0 \geq c_0^*$ the concentration of free monomer and all the cluster concentrations remain constant, which at first seems impossible. What these equations have predicted is, of course, simply that there is a saturation limit in the solution and that all molecules beyond the concentration c_0^* condense out of solution (i.e. from an all-interior phase which might be thought of as an infinite, no-surface aggregate). Thus our simple model has predicted a condensation type of phase transition.

The property of having a concentration c_0^* above which $x = 1$ and the distribution remaining constant results only if $y < 1$, which requires the surface to be unfavourable (expression (1.147)). Δg_s^0 can be written, as all free energies can be, in terms of an enthalpy and entropy

$$\Delta g_s^0 = \Delta h_s^0 - T\Delta s_s^0. \tag{1.151}$$

While Δg_s^0 is greater than zero mainly because Δh_s^0 is unfavourable, molecules on the surface are less restrained than those in the interior. If this is the case, then Δs_s^0 is positive. Since Δs_s^0 contributes to Δg_s^0 through the term $-T\Delta s_s^0$, it is possible for Δg_s^0 to become less than zero at high temperatures. When this happens y becomes greater than 1 and the sum in eqn (1.144) diverges at $x = 1$; hence $x = 1$ is no longer a solution, x simply approaching unity as c_0 becomes large. As we have already mentioned, if $y > 1$ the surface is favourable and the assumption of spherical clusters breaks down. However, it is clear that, if $y > 1$, there is no longer a condensation characteristic of the solution $x = 1$. Hence the temperature at which $y = 1$ corresponds to the loss of the condensation phase transition, or in other words our simple model exhibits a critical point.[9]

If the aggregates take the form of round platelets with a fixed thickness then the geometric parameters are

$$\text{area of platelet} \sim n \sim \pi r^2$$
$$\text{circumference of platelet} = 2\pi r \sim n^{\frac{1}{2}}. \tag{1.152}$$

Then, introducing free energies characteristic of the interior and the surface (edge of the platelet), we have, in analogy with eqn (1.141),

$$K_n = \exp(-an^{\frac{1}{2}}\Delta g_s^0/RT) \exp(-bn \, \Delta g_i^0/RT),$$

giving for eqn (1.108)

$$\sum_{n=1}^{\infty} y^{n^{\frac{1}{2}}} x^n = c_0, \tag{1.154}$$

with

$$y = \exp(-a\Delta g_s^0/RT),$$
$$x = [A_1] \exp(-b\Delta g_i^0/RT). \tag{1.155}$$

The analysis and qualitative properties of eqn (1.154) follow those of eqn (1.144).

In the formation of micelles there is a well-known feature which is not predicted by either the spherical or platelet models of aggregates discussed above. This feature is the presence of a critical micelle concentration (cmc, not to be confused with c_0^* of eqn (1.148)), which experimentally is as follows.[10] As we add monomer to a solution there is no tendency for aggregates to form to any large extent (for example, as shown by light-scattering) until we reach a characteristic concentration above which aggregates are very probable, as shown by the opalescence of the solution (we can see this effect by carefully diluting a cloudy soap solution, the clearing concentration corresponding to the critical micelle concentration). This feature is easy to predict if we characterize cluster

SIMPLE MULTIPLE EQUILIBRIA

formation by a single reaction

$$nA \rightleftharpoons A_n, \quad K_n. \tag{1.156}$$

We have the conservation of monomer

$$[A] + n[A_n] = c_0, \tag{1.157}$$

which using the law of mass action for the reaction in eqn (1.156) gives

$$[A] + n[A]^n K_n = c_0. \tag{1.158}$$

Defining

$$x = [A]K^n,$$
$$K = K_n^{1/(n-1)}, \tag{1.159}$$

eqn (1.159) reads

$$x + nx^n = Kc_0. \tag{1.160}$$

The terms x and nx^n are proportional to the probability of free monomer and monomer found in a cluster respectively. It is clear that

$$x < 1, \quad x \gg nx^n$$
$$\tag{1.161}$$
$$x > 1, \quad x \ll nx^n.$$

The larger n, the sharper the transition from free monomer to aggregate. Thus at $x = 1$ there is a sudden shift from monomer to aggregate. This corresponds to the critical micelle concentration or (eqn (1.160) with $x = 1$)

$$\text{cmc} = (1+n)/K \simeq n/K. \tag{1.162}$$

The reason for the sudden appearance of aggregates is that

eqn (1.156) allows only two species, monomer or large aggregate. In contrast, eqn (1.104) allows all stages of aggregation, which is characterized for all the models we have discussed (linear chain, spherical aggregate, platelet) by a cluster probability distribution similar to that shown in Fig. 1.3, that is, clusters of large size continuously become more probable, there being no sudden jump. Physically there is no good reason to restrict the system to a single aggregate size, as in eqn (1.156). On the other hand, it is equally clear that equations like (1.142) and (1.153), while reasonable for large spheres or platelets, might be unreasonable for very small clusters. For example, in a cluster of two, three, or four molecules, what is surface and what is interior? Thus we can amend eqns (1.142) and (1.153) to read

$$K_1 = 1$$

special formulation of K_n for $1 < n < n_0$ \hfill (1.163)

$$n \geq n_0, \quad K_n = \exp(-\Delta G_s(n)/RT)\,\exp(-\Delta G_i(n)/RT),$$

where n_0 is the aggregate size above which it makes sense to talk about spheres or platelets; the subscripts s and i refer to surface and interior respectively. If we assume that intermediate aggregates are very improbable, then we could set $K_n = 0$ for $1 < n < n_0$, in which case eqn (1.108) would become

$$[A_1] + \sum_{n=n_0}^{\infty} n[A_1]^n K_n = c_0 , \hfill (1.164)$$

which like eqn (1.157) would give rise to a critical micelle concentration, but unlike eqn (1.157) would allow aggregates of all sizes above $n = n_0$.

1.10. Some examples of binding equilibria

To conclude this chapter on systematic techniques to treat simple multiple equilibria we give two simple applications. Both involve the binding of small molecules to proteins, the first concerning the analysis of the states of binding as an

SIMPLE MULTIPLE EQUILIBRIA

intermediate step in determining the distance between binding sites on a protein, the second being the classic problem of the binding of O_2 to haemoglobin.

In the first example we have a protein which consists of two identical subunits each having an identical binding site to which various dye molecules can bind;[11] we designate this molecule by the symbol —S—S—. For a single dye species A the equilibria are

$$\begin{aligned} \text{—S—S—} + A &\rightleftharpoons \text{—S—S—A}, & K_A, \\ \text{—S—S—} + A &\rightleftharpoons \text{A—S—S—}, & K_A, \\ \text{—S—S—A} + A &\rightleftharpoons \text{A—S—S—A}, & K_A, \\ \text{A—S—S—} + A &\rightleftharpoons \text{A—S—S—A}, & K_A, \\ \text{—S—S—} + 2A &\rightleftharpoons \text{A—S—S—A}, & K_A^2. \end{aligned} \quad (1.165)$$

This model is identical with the dicarboxylic acids with independent carboxyl groups treated earlier. Hence

$$p(\text{—S—S—}) = 1/\xi,$$

$$p(\text{A—S—S—}) = p(\text{—S—S—A}) = [A]K_A/\xi, \quad (1.166)$$

$$p(\text{A—S—S—A}) = [A]^2 K_A^2/\xi,$$

$$\xi = 1 + 2[A]K_A + [A]^2 K_A^2 = (1+[A]K_A)^2.$$

The fraction ϑ of binding sites occupied is given simply by

$$\vartheta = \frac{[A]K_A}{1 + [A]K_A}. \quad (1.167)$$

From a study of ϑ as a function of [A] we can experimentally determine K_A. For another dye species B we have exactly the same equations with [A] and K_A replaced by [B] and K_B respectively, and from a separate binding study we can experimentally determine K_B. Now we wish to consider the case where we mix —S—S— with both A and B. Then the possible species and their contribution to ξ, the sum over all states of binding, are given below.

species	contribution to ξ
–S–S–	1
–S–S–A, A–S–S–	$[A] K_A$
–S–S–B, B–S–S–	$[B] K_B$
A–S–S–A	$[A]^2 K_A^2$
B–S–S–B	$[B]^2 K_B^2$
A–S–S–B, B–S–S–A	$[A][B] K_A K_B$.

Then ξ is given by

$$\xi = 1 + 2[A]K_A + 2[B]K_B + 2[A][B]K_A K_B + [A]^2 K_A^2 + [B]^2 K_B^2$$
$$= (1+[A]K_A+[B]K_B)^2. \qquad (1.168)$$

The quantity of interest in this problem is the concentration of A–S–S–B, since from the fluorescence properties of this species we can estimate the distance between the binding sites. From eqn (1.165) the probability of this mixed species is simply

$$p(\text{A–S–S–B} + \text{B–S–S–A}) = 2[A][B]K_A K_B/\xi . \qquad (1.169)$$

The second example is the binding of O_2 to haemoglobin. Haemoglobin consists of four subunits each of which contains a haem group capable of binding O_2. The equilibria for this system are

$$\begin{aligned}
\text{Hm} &\rightleftharpoons \text{Hm} & K_0 &= 1 \\
\text{Hm} + O_2 &\rightleftharpoons \text{Hm}(O_2), & K_1 & \\
\text{Hm} + 2O_2 &\rightleftharpoons \text{Hm}(O_2)_2, & K_2 & \\
\text{Hm} + 3O_2 &\rightleftharpoons \text{Hm}(O_2)_3, & K_3 & \\
\text{Hm} + 4O_2 &\rightleftharpoons \text{Hm}(O_2)_4, & K_4 &
\end{aligned} \qquad (1.170)$$

SIMPLE MULTIPLE EQUILIBRIA

A general relation for the average fraction of sites occupied is given immediately as (see eqn (1.64))

$$\vartheta = \tfrac{1}{4}\left\{\frac{[O_2]K_1 + 2[O_2]^2 K_2 + 3[O_2]^3 K_3 + 4[O_2]^4 K_4}{1 + [O_2]K_1 + [O_2]^2 K_2 + [O_2]^3 K_3 + [O_2]^4 K_4}\right\}. \quad (1.171)$$

(The concentration of O_2 in solution is directly proportional to the partial pressure of O_2 above the solution.) If the binding sites are independent and equivalent then

$$\vartheta = \frac{[O_2]K}{1 + [O_2]K}, \quad (1.172)$$

$\xi = (1+[O_2]K)$, $K_1 = 4K$, $K_2 = 6K^2$, $K_3 = 4K^3$, $K_4 = K^4$,

where K is the binding constant to a single site. If eqn (1.172) were true, then ϑ versus $[O_2]$ would have the form shown by the dashed curve in Fig. 1.12. Experimentally the solid

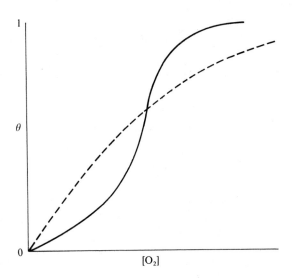

FIG. 1.12. Schematic illustration of the binding of O_2 to haemoglobin. The dotted curve shows the behaviour that would obtain if the binding to the four sites were independent; the solid curve shows the actual nature of the binding illustrating that the binding becomes easier in partially saturated molecules.

curve is found, which means that the sites are not independent.[12] Since there is a large distance between the sites, there can be no physical interaction between the bound oxygens, the rationale of the non-independence being that the binding of one oxygen induces a conformational change in the subunit which alters the binding constants of other sites. The constant K_1 is easily determined since

$$\left(\frac{\partial \vartheta}{\partial [O_2]}\right)_{[O_2]=0} = K_1, \qquad (1.173)$$

i.e. the slope of ϑ versus O_2 at zero O_2 concentration gives the first binding constant. The sigmoidal shape indicates that K_2, K_3, or K_4 must be larger than the constants predicted for independent sites (eqn (1.172)).

An empirical relation which reproduces the actual binding of O_2 by haemoglobin is known as the Hill-Barcroft equation[13]

$$\vartheta = \frac{[O_2]^n K}{1 + [O_2]^n K}, \qquad (1.174)$$

where the value of n that gives the best fit to the curve for normal blood is approximately 2.5. Clearly eqn (1.174) is an approximation to the more general eqn (1.171); the value $n = 2.5$ indicates that K_2 and K_3 probably dominate in eqn (1.171). Eqn (1.174) is similar to the results following eqn (1.156) in our discussion of micelle formation.

The oxygenation of haemoglobin is a simple example of a cooperative process: the binding of a single O_2 molecule to haemoglobin is difficult, but once the binding is nucleated, successive oxygen molecules bind easier. Fig. 1.12 shows that the cooperative nature of the binding produces a characteristic sigmoidal curve which is qualitatively different from the results predicted by binding to independent sites.

1.11. Summary

In this chapter I have tried to show that simple multiple equilibria are easily treated if we proceed exactly as we do for the case of equilibria between quantum-mechanical energy

SIMPLE MULTIPLE EQUILIBRIA

levels. If we have N species, we need only consider N free energy levels or, equivalently, write all possible reactions with respect to a single species. Then the probability distribution for the possible species is immediately given by

$$p(\text{species}) = \frac{\exp(-\text{free energy(species)}/RT)}{\sum_{\text{species}} \exp(-\text{free energy(species}/RT)}. \quad (1.175)$$

With a little practice with this technique we can write down the solution to rather complicated multiple-equilibria problems immediately by inspection. Without this technique it is easy to find ourselves in a circular maze of interwoven equations.

Notes and References

1. See either of the books by Hill in ref. 4, Chapter 2.
2. The origin of this relation is discussed in detail in Chapter 2.
3. Deviations from the law of mass action for this reaction are discussed in Chapter 2.
4. The quantity ξ is the grand partition function for a single molecule; see section 2.1.8.
5. For a tabulation of many pK values see: *Handbook of chemistry and physics*, Chemical Rubber Publishing Co., Cleveland, Ohio (1974).
6. See ref. 9 for many references to the physical cluster model.
7. Schachman, H.K., Alder, J., Radding, C.M., Lehman, I.R., and Kornberg, A., *J. biol. Chem.* **235**, 3242 (1960).
8. For the phosphorylase catalysed polymerization of glucose 1-phosphate on glycogen primer see: Cori, G.T., Swanson, M.A., and Cori, C.F., *Fedn Proc. Fedn Am. Socs exp. Biol.* **4**, 234 (1945); Trevelyan, W.E., Mann, P.F.E., and Harrison, J.S., *Archs Biochem. Biophys.* **39**, 419 (1952); Hestrin, S., *J. biol. Chem.* **179**, 943 (1949).
9. Fisher, M., *Physics* **3**, 255 (1967).
10. See, for example: Poland, D. and Scheraga, H.A., *J. phys. Chem.* **69**, 2431 (1965).
11. Conrad, R.H., Hertz, J.R., and Brand, L., *Biochemistry* **9**, 1540 (1970).
12. Bock, A.V., Field, H. Jr., and Adair, G.S., *J. biol. Chem.* **59**, 353 (1924).
13. Barcroft, J., *The respiratory function of the blood*, Cambridge University Press (1914); *The respiratory function of the blood, II. Haemoglobin*, Cambridge University Press (1928).

2
BACKGROUND: PARTITION FUNCTIONS AND INTERMOLECULAR FORCES

In the last chapter I presented a general method for treating simple multiple equilibria. To understand and use the technique presented requires only a minimum of background in physical chemistry: simple Boltzmann distribution, equilibrium constants, and the law of mass action. In subsequent chapters we will extend the technique to describe cooperative conformational transitions in biological macromolecules, an extension which involves little more in principle than techniques for evaluating the sums appearing in eqn (1.95) (of course our main interest is not technique but what an applied technique can tell us about molecules). Since the approach of the last chapter emphasized sums over free energy levels involving partition function-like expressions, at this point in our discussion we want to explore the nature of these sums in more detail. Our chief aim will be to establish a link between the formalism outlined in the last chapter and the standard models treated in any text on statistical mechanics (ideal gas, harmonic oscillator, imperfect gas, etc.). While the reader can follow subsequent chapters using only the background in Chapter 1, the present chapter is not intended as an excuse for presenting standard equations from statistical mechanics; rather it will serve as a link between empirical parameters such as equilibrium constants which appear in the formalism of Chapter 1 and parameters that describe the interactions between molecules. I do not promise that the contents of this chapter will enable the reader to calculate in exact detail all (or even some!) of the parameters required to treat a biological macromolecule in water. What it will do is enable the reader to understand why a particular equilibrium constant or free energy is large or small, is dominated by entropy or energy, and what the important intermolecular energies are. And this insight is very important in the construction of realistic models (such as the model of spherical aggregates treated in the last chapter) which requires answers to questions like: What are the important states or free energies? Which states can be ignored or

BACKGROUND: PARTITION FUNCTIONS

successfully approximated? Under what conditions do the approximations hold or break down?

Our discussion will centre around two topics — partition functions and intermoleculer forces — which we discuss in turn.

2.1. Partition functions

Statistical mechanics forms the bridge between the properties of single atoms and molecules (as described by quantum mechanics) and the observable properties of bulk matter (as described by thermodynamics). While the mathematics of quantum mechanics becomes very difficult for any system more complicated than the hydrogen atom, most chemists have a sound understanding of the concepts of quantum mechanics such as wavefunctions, electron density, and energy levels. After all, although it may be impossible to solve exactly, the formulation of a Schrödinger equation for a multi-electron system is conceptually quite simple: there are terms for the kinetic energy and terms for the potential energy of each electron and nucleus (interacting with each other). Thermodynamics, on the other hand, is a subject where the mathematical component is very simple but where the conceptual component is very hazy for most students. For example, all of the quantities U, H, G, F, pV, μ, and TS have the dimensions of energy. How are these related to the simple ideas of kinetic and potential energy (which are preserved in quantum mechanics from Newtonian mechanics)? Since our study of multiple equilibria of necessity will involve these thermodynamic quantities, it is essential that they be well understood, both with respect to what they are and what they are not. So we are going to preface our remarks on statistical mechanics with a brief intuitive excursion into thermodynamics. By intuitive it is meant that we will freely use ideas that properly belong to statistical mechanics when they clarify the meaning of a thermodynamic variable. Recall that in thermodynamics proper no reference is made to a molecular theory of matter; for our purpose here we want to emphasize the molecular interpretation, though in so doing we will somewhat corrupt the beautiful logic of thermodynamics where far-reaching conclusions are drawn solely from a few observations on bulk matter.

2.1.1. Thermodynamics

A collection of atoms and molecules is just like any other system of particles in physics in that they have both kinetic and potential energy. The kinetic energy of a point particle of mass m is

$$\text{Kinetic energy} = \tfrac{1}{2}mv^2,$$

where v is the velocity; or in terms of the momentum, $p = mv$,

$$\text{Kinetic energy} = p^2/2m. \qquad (2.1)$$

Since atoms and molecules are made up of protons and electrons, the potential energy all ultimately comes simply from Coulomb's law,

$$\text{potential energy} = q_1 q_2 / r, \qquad (2.2)$$

where r is the distance between the charges q_1 and q_2. In section 2.2 of this chapter we will note that the potential energy between molecules takes many different forms (functions of r); however, they all have their origin in the simple Coulombic potential of eqn (2.2).

The simplest and most basic thermodynamic function is the internal energy, U, where

$$U = \begin{pmatrix} \text{average kinetic} \\ \text{energy of the} \\ \text{molecules of} \\ \text{a system} \end{pmatrix} + \begin{pmatrix} \text{average potential} \\ \text{energy of the} \\ \text{molecules of} \\ \text{a system} \end{pmatrix}. \qquad (2.3)$$

Thus U is an extension of the notions of kinetic energy and potential energy to a many-body system. All of thermodynamics[1] is contained in the equation for the total differential of U (for the present, at constant number of particles)

$$dU = TdS + \sum_i X_i dx_i, \qquad (2.4)$$

where S is the entropy and

X_i = generalized force,

x_i = generalized displacement, (2.5)

$X_i \mathrm{d}x_i$ = generalized differential of work.

Some of the common generalized forces and their respective displacements are:

X_i	$\mathrm{d}x_i$	nature of work
P	$-\mathrm{d}V$	pressure-volume
γ	$\mathrm{d}A$	surface-tension-area
f	$\mathrm{d}L$	force-length.

(2.6)

Any or all of the $X_i \mathrm{d}x_i$ terms may be present depending on the nature of the system being studied. The commonest work term for fluids is $-p\mathrm{d}V$. Thus eqn (2.4) is often written as

$$\mathrm{d}U = T\mathrm{d}S - p\mathrm{d}V. \qquad (2.7)$$

While we will use the special case of eqn (2.4) given in eqn (2.7), we must remember that eqn (2.4) is the general starting point for developing all thermodynamic relations. In biophysical applications surface-tension-area work (for films, membranes) and force-length work (for fibres) are important work terms.

Eqn (2.4) contains the results of both the first and second laws of thermodynamics. The first law postulates the existence of the function of state U with the property

$$\Delta U = \text{(heat added to system)} + \text{(work done on system)}. \qquad (2.8)$$

Eqn (2.8) states that the sum of the kinetic energy and the potential energy of the molecules of a given system can be increased by adding heat to the system (molecular mechanism of transferring energy: collisions with other molecules, absorption of radiation) or by doing work on the system (macroscopic mechanism of transferring energy: compression, stirring, etc.).

Since what we are adding in both mechanisms is energy, the molecules obviously cannot tell where the energy comes from (heat or work). Thus U is a state function (the molecules of a system have a definite average kinetic and potential energy) but heat and work are not (they are just two ways of adding the same thing). The second law postulates the existence of a second state function, the entropy S, defined as

$$dS = dq_{rev}/T, \qquad (2.9)$$

where dq_{rev} is the heat added to the system in a reversible manner. Note that TdS is the heat added to and $-pdV$ the work done on a system if the change is done reversibly; if the change is not reversible, eqn (2.7) still holds since it involves state functions only (a net change depending on the initial and final states only and not on the path between them), but in the irreversible case TdS and $-pdV$ are not equal to increments of heat and work respectively. Thus if we specify the conditions under which energy is added to a system, we can use the addition of heat or work to experimentally measure S and U. In particular, introducing the heat capacity at constant volume

$$C_V = \left(\frac{\partial U}{\partial T}\right)_V = \frac{dq}{dT}, \qquad (2.10)$$

we have the practical relations

$$\Delta U = \int_{T_1}^{T_2} C_V(T) dT,$$

$$\Delta S = \int_{T_1}^{T_2} \{C_V(T)/T\} dT. \qquad (2.11)$$

Heat capacities are determined by adding a known amount of heat to a system under specific conditions (for example, constant volume) and measuring the resulting ΔT; the ratio, in the limit of zero ΔT, is the heat capacity.

While for gases it is a simple matter to keep the volume constant, for condensed phases this is difficult, hence constant pressure (the system open to the atmosphere) is the common condition. With constant pressure

$$dq = dU + pdV = d(U+pV) = dH,$$

$$H = U + pV,$$

$$C_p = \left(\frac{\partial H}{\partial T}\right)_p = \frac{dq}{dT}, \quad (2.12)$$

$$\Delta H = \int_{T_1}^{T_2} C_p(T) dT.$$

The thermodynamic function H, the enthalpy, is simply a definition to avoid writing $U + pV$. Unfortunately many books, after introducing H, immediately identify changes in this quantity (e.g., for a chemical reaction) with a difference in chemical bond energy. From eqns (2.12) and (2.3)

$$H = \begin{Bmatrix} \text{average kinetic} \\ \text{energy of the} \\ \text{molecules of} \\ \text{a system} \end{Bmatrix} + \begin{Bmatrix} \text{average potential} \\ \text{energy of the} \\ \text{molecules of} \\ \text{a system} \end{Bmatrix} + pV. \quad (2.13)$$

Clearly both U and H contain valuable information about the potential energy of molecules (a measure of the amount and nature of the bonding in the system). Strictly

$$\langle E_{pot} \rangle = U - \langle E_{kin} \rangle,$$
$$\langle E_{pot} \rangle = H - \langle E_{kin} \rangle - pV. \quad (2.14)$$

As we will review shortly, both the average kinetic energy and pV are of the order of RT, which at room temperature is of the order of 2.5 kJ mol^{-1}. Thus if the source of the potential energy is chemical bonds with energies of the order of -8.36 to -418 kJ mol^{-1}, both $\langle E_{kin} \rangle$ and pV are small corrections. However, if we are talking about weak energies such as hydrogen bonds (~ -20.9 kJ mol^{-1}) or van der Waals interactions (~ -12.5 kJ mol^{-1}) then the $\langle E_{kin} \rangle$ and pV terms are not negligible. It is very important to remember that eqn (2.14) is the correct relation of the potential energy of molecules to H and U (eqn (2.14), of course, is a statistical-mechanical result and is not part of thermodynamics proper, which does not necessarily admit the existence of molecules).

As I have already mentioned, eqn (2.7) (or eqn (2.4)) is *the* equation of thermodynamics, containing the results of both first and second laws. Together with the important inequality of Clausius,

$$\text{(isolated system)} \quad dS \geqslant 0, \tag{2.15}$$

eqn (2.7) is the source of all the other thermodynamic functions and relations. Of paramount importance are the conditions for equilibrium under various conditions. These are (eqn (2.7)) under the special conditions listed):

Variables held constant	condition for equilibrium	characteristic thermodynamic function (thermodynamic potential)
S,V	$dU = 0$ ($U \to$ minimum)	U (internal energy)
U,V	$dS = 0$ ($S \to$ maximum)	S (entropy)
V,T	$d(U-TS) = 0$ ($F = U-TS \to$ minimum)	F (Helmholtz free energy)
p,T	$d(U-TS+pV) = 0$ ($G = U-TS+pV = H-TS = F+pV \to$ minimum)	G (Gibbs free energy)

(2.16)

The important features of the above tabulation are: all of the results come from eqn (2.7) with eqn (2.15); the thermodynamic potential (characteristic function) depends on the conditions (variables held constant); and F and G are definitions to avoid writing $U - TS$ and $U - TS + pV$, respectively.

Since the conditions of constant T and V (for gases) and constant p and V (for liquids and solids) are the usual experimental conditions, F and G are the thermodynamic potentials most commonly encountered in applications. From the definitions of F and G (eqn (2.16) and eqn (2.7)) we obtain

$$dF = -SdT - pdV = \left(\frac{\partial F}{\partial T}\right)_V dT + \left(\frac{\partial F}{\partial V}\right)_T dV,$$

$$dG = -SdT + Vdp = \left(\frac{\partial G}{\partial T}\right)_p dT + \left(\frac{\partial G}{\partial p}\right)_G dp, \quad (2.17)$$

where the right-hand sides of eqn (2.17) are mathematical relations for the functions $F(T,V)$ and $G(T,p)$ with the independent variables (T and V) and (T and p), respectively, the middle expressions being, of course, thermodynamic relations (results of the combined first and second law equation, eqn (2.7)). From eqn (2.17) we have the identifications

$$S = -\left(\frac{\partial F}{\partial T}\right)_V, \quad p = -\left(\frac{\partial F}{\partial V}\right)_T,$$

$$S = -\left(\frac{\partial G}{\partial T}\right)_V, \quad V = \left(\frac{\partial G}{\partial p}\right)_T. \quad (2.18)$$

Combining eqn (2.18) for S and the definitions of F and G we obtain

$$F = U - TS = U + T\left(\frac{\partial F}{\partial T}\right)_V,$$

or

$$G = H - TS = H + T\left(\frac{\partial G}{\partial T}\right)_p,$$

$$U = \left(\frac{\partial F/T}{\partial 1/T}\right)_V = -T^2\left(\frac{\partial F/T}{\partial T}\right)_V, \quad (2.19)$$

$$H = \left(\frac{\partial G/T}{\partial 1/T}\right)_p = -T^2\left(\frac{\partial G/T}{\partial T}\right)_p.$$

Using the definitions of C_V and C_p we further have

$$C_V = -T\left(\frac{\partial^2 F}{\partial T^2}\right)_V,$$

$$C_p = -T\left(\frac{\partial^2 G}{\partial T^2}\right)_p. \quad (2.20)$$

Summarizing eqns (2.18)–(2.20) we have the basic set of thermodynamic relations:

independent variables V and T thermodynamic potential = $F(V,T)$	independent variables p and T thermodynamic potential = $G(p,T)$
$S = -\left(\dfrac{\partial F}{\partial T}\right)_V$	$S = -\left(\dfrac{\partial G}{\partial T}\right)_p$
$p = -\left(\dfrac{\partial F}{\partial V}\right)_T$	$V = \left(\dfrac{\partial G}{\partial p}\right)_T$
$U = \left(\dfrac{\partial F/T}{\partial 1/T}\right)_V$	$H = \left(\dfrac{\partial G/T}{\partial 1/T}\right)_p$
$C_V = -T\left(\dfrac{\partial^2 F}{\partial T^2}\right)_V$	$C_p = -T\left(\dfrac{\partial^2 G}{\partial T^2}\right)_p$
$H = U + pV$	$U = H - pV$
$G = F + pV$	$F = G - pV$

(2.21)

The importance of tabulation (2.21) is that if either $F(V,T)$ or $G(p,T)$ is known, all the other thermodynamic functions are given as appropriate derivatives of F or G.

If we recall the basic relations of statics (physics of mechanical equilibrium):

$$\text{force} = -\text{ gradient of potential energy}$$
$$\text{at equilibrium: net force} = 0 \qquad (2.22)$$
$$\text{potential energy} \rightarrow \text{minimum,}$$

then we see that for bulk matter F of G (or U or S depending on the independent variables) plays the role of the potential energy for equilibrium in simple mechanical systems:

force =	− gradient of thermodynamic potential	with respect to
S	F (at constant V)	T
p	F (at constant T)	V
U	$-F/T$ (at constant V)	$1/T$
S	G (at constant p)	T
V	$-G$ (at constant T)	p
H	$-G/T$ (at constant p)	$1/T$

(2.23)

at equilibrium: $F \to$ minimum (at constant V and T)

$G \to$ minimum (at constant p and T).

So far we have been discussing systems with a constant number of particles. If the number and type of particles (N_i = number of particles of type i) are variable, then eqns (2.7) and (2.17) become

$$dU = TdS - pdV + \sum_i \mu_i dN_i,$$

$$dF = -SdT - pdV + \sum_i \mu_i dN_i, \qquad (2.24)$$

$$dG = -SdT + Vdp + \sum_i \mu_i dN_i,$$

where

$$\mu_i = \left(\frac{\partial U}{\partial N_i}\right)_{V,S,N_j} = \left(\frac{\partial F}{\partial N_i}\right)_{T,V,N_j} = \left(\frac{\partial G}{\partial N_i}\right)_{T,p,N_j}$$

(2.25)

(all N_j constant, $j \neq i$).

The quantity μ_i, the chemical potential, plays a central role in chemical equilibrium. At constant S and V, or T and V, or T and p, eqn (2.24) give respectively (the same result)

at equilibrium: $\sum_i \mu_i dN_i = 0.$ (2.26)

Eqn (2.26) is the general thermodynamic relation for chemical equilibrium.

Eqn (2.26) can be simplified for systems undergoing chemical reactions by using the fact that the stoichiometry of the reactions reduces the number of independent variables dN_i. A general chemical reaction

$$aA + bB + \ldots \rightleftharpoons cC + dD + \ldots$$

can be rewritten as

$$(cC + dD + \ldots) - (aA + bB + \ldots) = 0, \qquad (2.27)$$

which emphasizes the fact that eqn (2.27) is a conservation relation. Introducing[2] the stoichiometric coefficients ν_i

$$\begin{aligned} \nu_C &= c, & \nu_A &= -a, \\ \nu_D &= d, & \nu_B &= -b, \end{aligned} \qquad (2.28)$$

then we can interpret eqn (2.28) as simply the conservation of mass

$$\sum_i \nu_i m_i = 0, \qquad (2.29)$$

the m_i being the masses of the respective species. For eqn (2.27) there is only one independent concentration variable. Defining a progress variable ξ as the single required variable then

$$dN_i = \nu_i d\xi. \qquad (2.30)$$

Substituting eqn (2.30) into eqn (2.26) we have

$$\text{at equilibrium:} \quad (\sum_i \mu_i \nu_i) d\xi = 0. \qquad (2.31)$$

At equilibrium a differential change $d\xi$ must give eqn (2.31); hence

$$\text{at equilibrium:} \quad \sum_i \mu_i \nu_i = 0. \qquad (2.32)$$

Eqn (2.32) is a general condition for chemical equilibrium; note in particular that eqn (2.32) applies to any system and that no restricting assumptions have been made.

Pursuing the analogy between equilibrium in simple mechanical systems (statics) and bulk matter (thermodynamics) outlined in eqns (2.22) and (2.23) we have

simple mechanical systems	bulk matter
force = $-\begin{pmatrix}\text{gradient of}\\ \text{potential}\\ \text{energy}\end{pmatrix}$	chemical potential = $\begin{pmatrix}\text{gradient of}\\ \text{thermodynamic}\\ \text{potential}\end{pmatrix}$
at equilibrium: $\sum$ forces = 0	$\sum\begin{pmatrix}\text{chemical}\\ \text{potentials}\end{pmatrix} = 0$

(2.33)

Note that μ, the chemical potential, actually plays the role of the analogue of a force (chemical gradient or force would be a more descriptive name), F (or G, or U, or S depending on the appropriate independent variables) playing the role of the analogue of the potential energy.

For equilibrium between two phases

$$\text{phase I} \rightleftharpoons \text{phase II} \tag{2.34}$$

we have $\mu_{II} - \mu_{I} = 0$ or

$$\mu_{II} = \mu_{I}. \tag{2.35}$$

The form of eqn (2.35), the equality of chemical potentials, is often given as the criterion for equilibrium, though eqn (2.32) is a more general relation applying to any chemical reaction.

Since it plays such an important role in equilibrium, we should have a clear notion of what the chemical potential is. This is perhaps most easily achieved by considering the experiment necessary to measure it. For a multicomponent system we keep the number of particles of all the components except one constant and measure the free energy as a function of the concentration of one component. The slope of the variation of the free energy as a function of the number of particles of the one component at a particular value of the number of particles of that component is the chemical potential. As the number of particles of one component is changed, the probability of this component interacting with all the other components in the system obviously changes. Thus μ_i in general is a function of the concentrations of all the other components in the system.

Mathematics alone gives a further important relation involving the chemical potential. For $F(V,T,N_i)$ and $G(p,T,N_i)$ the independent variables may be classified as either extensive (V,N_i) or intensive (p,T). Since

$$F(\alpha V, T, \alpha N_i) = \alpha F(V,T,N_i),$$
$$G(p, T, \alpha N_i) = \alpha G(p,T,N_i),$$
(2.36)

which simply states that if we multiply all the extensive variables by α we multiply the free energy by α (e.g. 2 litres of water has twice the free energy of 1 litre of water at the same p and T). Euler's theorem for homogeneous functions tells us that functions with the property of eqn (2.36) can be written as a linear combination of the extensive variables

$$F = \sum_i \mu_i N_i - pV,$$
$$G = \sum_i \mu_i N_i.$$
(2.37)

Eqns (2.37) are completely general results; they are also deceptive in their simplicity. Recall that μ_i in general is a function of the concentrations of all the other components in the system. Note that eqn (2.37) does not say that

$$\begin{pmatrix}\text{for independent}\\ \text{systems only}\end{pmatrix} \quad F = \sum_i F_i + (\sum_i p_i)V,$$
$$G = \sum_i G_i.$$
(2.38)

As we have indicated above, eqn (2.38) are special relations that only apply to independent components (ideal systems). For a one-component system eqn (2.37) gives

$$G = N\mu$$
(2.39)

or

$$\mu = \partial G/\partial N = G/N. \qquad (2.40)$$

Eqn (2.40) is a general result for a one-component system: for a one-component system μ is the Gibbs free energy per particle (or per mole depending on the units one is using for energy). Thus while eqn (2.25) give μ as appropriate derivatives of U, F, or G, eqn (2.40) shows that μ has a special relation with the Gibbs free energy.

With the exception of eqn (2.38), all the thermodynamic relations I have given are general results applying to any system. We now consider an important approximate result. For dilute gases or solutions we find (we will explore the origins of this relation shortly)

$$\begin{pmatrix}\text{ideal gases}\\ \text{or solutions}\end{pmatrix} \qquad \mu_i(T,\rho_i) = \mu_i^0(T) + RT\ln\rho_i, \qquad (2.41)$$

where

$$\rho_i = N_i/V. \qquad (2.42)$$

is the density (or concentration) and μ_i^0 is a temperature-dependent term characteristic of the ith species (independent of the density). In particular note that μ_i in eqn (2.41) does not depend on the concentration of the other components. Taking

$$(\text{general relation}) \quad \sum_i \mu_i \nu_i = 0, \qquad (2.32)$$

$$\begin{pmatrix}\text{special form of } \mu_i \text{ for}\\ \text{ideal gases or solutions}\end{pmatrix} \mu_i = \mu_i^0 + RT\ln\rho_i, \qquad (2.41)$$

the following successive steps

$$\sum_i \mu_i \nu_i = \sum_i \mu_i^0 \nu_i + RT \sum_i \nu_i \ln\rho_i$$

$$= \sum_i \mu_i^0 \nu_i + RT \sum_i \ln\rho_i^{\nu_i}$$

$$= \sum_i \mu_i^0 \nu_i + RT \prod_i \ln \rho_i^{\nu_i}$$

yield

$$\prod_i \rho_i^{\nu_i} = \exp(-\sum_i \mu_i^0 \nu_i / RT) \qquad (2.43)$$

Since the μ_i^0 are at most functions of temperature

$$\exp(-\sum_i \mu_i^0 \nu_i / RT) = K(T) \qquad (2.44)$$

or

$$\prod_i \rho_i^{\nu_i} = K(T). \qquad (2.45)$$

Eqn (2.43) or (2.45) is of course the law of mass action, which we used extensively in Chapter 1. For example, for the reaction

$$aA + bB \rightleftharpoons cC + dD \qquad (2.46)$$

eqn (2.43) gives

$$\frac{[C]^c [D]^d}{[A]^a [B]^b} = \exp(-\Delta \mu^0 / RT), \qquad (2.47)$$

$$\Delta \mu^0 = c\mu_C^0 + d\mu_D^0 - a\mu_A^0 - b\mu_B^0$$

(recall that from eqn (2.28) some of the ν_i are negative).

We cannot emphasize enough that while eqn (2.32) is a general thermodynamic relation, eqn (2.43) (the law of mass action) is not, the form of eqn (2.43) coming from the specific (approximate) form of eqn (2.41) that only applies to ideal (or, in practical terms, very dilute) systems.

Since μ_i of eqn (2.41) is independent of the other components in the system

$$G = \Sigma G_i,$$

$\begin{pmatrix}\text{ideal gases}\\\text{or solutions}\end{pmatrix}$
$$\mu_i = G_i/N_i = g_i,$$
$$g_i = g_i^0 + RT\ln\rho_i,$$
$$g_i^0 = \mu_i^0.$$
(2.48)

The decoration on g_i^0 has the meaning: the subscript i refers to the species type while the superscript 0 indicates a characteristic value independent of the concentration (μ_i, μ_i^0, and g_i^0 are intensive quantities; G and G_i are extensive). The quantity $\mu_i^0 = g_i^0$ is simply a number characteristic of species i; eqn (2.41) tempts us to say that μ_i^0 or g_i^0 is the chemical potential or molar Gibbs free energy at unit concentration (hence the term standard free energy or standard chemical potential, the standard being unit concentration which, of course, depends upon the units one uses for concentration). This temptation should be avoided since eqn (2.41) only applies in the limit of low density (i.e. at unit concentration, which in common units would be 1 M, μ_i is not equal to μ_i^0 since eqn (2.41) no longer holds accurately). In fact eqn (2.41) is best interpreted as a limiting relation

$$\lim_{\rho_i \to 0} \begin{cases} \mu_i \to \mu_i^0 + RT\ln\rho_i \\ \mu_i^0 \to (\mu_i - RT\ln\rho_i). \end{cases} \quad (2.49)$$

In Chapter 1 we have used the symbol g_i^0, since the free energy is a more familiar concept than the chemical potential; the chemical potential is the more basic quantity in chemical equilibrium.

Since eqns (2.41) and (2.43) are extraordinarily useful approximations (all of Chapter 1 was based on these relations) it is of paramount importance that we understand the conditions under which they apply. In section 2.2 we will explore effects which cause deviations from these relations. But first we must briefly review the basic relations between thermodynamics and molecular properties, the science of this relationship, of course, being statistical mechanics. A familiarity with the basic relations of statistical mechanics, together with a knowledge of intermolecular forces (given in section 2.2 of this chapter), will enable us to understand the factors influencing the value of a particular μ_i^0 and what the molecular properties and conditions are that lead to deviations from the law of mass action.

2.1.2. Statistical mechanics: the molecular interpretation of thermodynamics

Thermodynamics deals with energy in profusion: U, H, G, G, TS, pV, and μ all have the dimensions of energy. Classical mechanics gives only two basic kinds of energy: kinetic and potential. And for systems like atoms or planets there are only two important potential energies,

$$\varphi(r) = -\frac{m_1 m_2}{r} \quad \text{Newton's law of gravitation,}$$

$$\varphi(r) = \frac{q_1 q_2}{r} \quad \text{Coulomb's law,} \quad (2.50)$$

which are strikingly similar with the important difference that $q_1 q_2$ can be positive or negative depending on whether the charges q are like or unlike. Other potentials, like the gravitational potential near the surface of the earth or Hooke's law for the displacement of a spring from its equilibrium position,

(gravitational potential near the surface of the earth) $\quad \varphi(h) = mgh$, (2.51a)

Hooke's law $\quad \varphi(\Delta x) = \tfrac{1}{2} K (\Delta x)^2$, (2.51b)

where Δx is the displacement from the equilibrium position, x_0, of the spring ($\Delta x = x - x_0$) and h is the distance of an object of mass m above the surface of the earth, are not basic potentials but mathematical approximations. Thus any potential can be expanded about a point $x = x_0$ in a Taylor series

$$\varphi(x) = \varphi_{x=x_0} + \left(\frac{\partial \varphi}{\partial x}\right)_{x=x_0}(x-x_0) + \tfrac{1}{2}\left(\frac{\partial^2 \varphi}{\partial x^2}\right)_{x=x_0}(x-x_0)^2 +$$
$$+ \frac{1}{3!}\left(\frac{\partial^3 \varphi}{\partial x^3}\right)_{x=x_0}(x-x_0) + \cdots \quad (2.52)$$

If we are expanding about a minimum in φ, then $(\partial \varphi/\partial x)_{x=x_0}$ is zero; if we take the zero of the scale of potential energy at the point $x = x_0$, then truncation of eqn (2.25) gives Hooke's law. Truncation of eqn (2.52) at the term linear in $h = x - x_0$ gives eqn (2.51a). For a spring, which is made of metal composed of atoms containing electrons and protons, the potential of eqn (2.52) (approximated by eqn (2.51b)) ultimately comes from a complex application (via the Schrödinger equation for the metal) of Coulomb's law.

In quantum mechanics[3] the notions of kinetic and potential energy are preserved in the formulation of the Hamiltonian (which is the basic starting point of classical mechanics as well):

$$\text{Hamiltonian} = \mathcal{H} = \text{total energy} \quad (2.53)$$
$$= \Sigma\binom{\text{kinetic-energy}}{\text{terms}} + \Sigma\binom{\text{potential-energy}}{\text{terms}}$$

where the kinetic-energy and potential-energy terms are of the form (see eqns (2.50), (2.1), and (2.2))

$$E_{\text{kin}} = p^2/2m,$$
$$E_{\text{pot}} = q_1 q_2/r. \quad (2.54)$$

The next step is to replace the momentum by a differential operator ($p_x = i\hbar\, \partial/\partial x$, p_x being the x component of the momentum). The time-independent Schrödinger equation then reads

$$\mathcal{H}\psi_i = \varepsilon_i \psi_i, \quad (2.55)$$

where $\mathcal{H}$ is a differential operator (through the momentum terms in eqn (2.53)), ψ_i are the characteristic wavefunctions, and the ε_i are the allowed energy levels. The point I wish to make here is that the starting point for obtaining the ε_i is the formulation of the Hamiltonian (eqn (2.53)) which is composed of kinetic- and potential-energy terms. In general there are an infinite number of allowed energy levels resulting from a solution of eqn (2.55). Quantum mechanics tells us nothing about the occupation of these levels; this question lies in the realm of statistical mechanics.

I have already quoted one of the central results of statistical mechanics, namely the Boltzmann distribution, which gives the probability of being in energy state ε_i as (eqn (1.13))

$$p_i = \exp(-\varepsilon_i/kT)/q,$$

$$q = \sum_i \exp(-\varepsilon_i/kT), \tag{2.56}$$

the quantity q (not to be confused with the charge in eqn (2.50)) being the partition function or sum over states. In eqn (2.56) and in the rest of section 2.1 we use k instead of R; this is necessary since we will soon be writing partition functions for N particles, N being of the order of Avogadro's number. In Chapter 1 we have examined the behaviour of the occupation probabilities p_i for several special cases.

The average energy of the states ε_i is by definition

$$\langle \varepsilon \rangle = \frac{\Sigma \varepsilon_i \exp(-\varepsilon_i/kT)}{\Sigma \exp(-\varepsilon_i/kT)} = -\frac{\partial q}{\partial (1/kT)}/q$$

$$= -k\frac{\partial \ln q}{\partial (1/T)}. \tag{2.57}$$

Since $\langle \varepsilon \rangle$ is just the average kinetic and potential energy, we can identify $\langle \varepsilon \rangle$ with the internal energy for the system characterized by the ε_i

$$u = \langle \varepsilon \rangle = -k\frac{\partial \ln q}{\partial (1/T)}. \tag{2.58}$$

From eqn (2.19) we have a relation for u in terms of the Helmholtz free energy which we write (here in units of energy per molecule, denoted by the lower-case symbols u, f, etc.)

$$u = \frac{\partial (f/T)}{\partial (1/T)}. \tag{2.59}$$

Equating the two expressions for u, eqn (2.58) and (2.59), gives

$$\frac{\partial (f/T)}{\partial (1/T)} = -k\frac{\partial \ln q}{\partial (1/T)}, \tag{2.60}$$

which suggests the identification

$$f = -kT\ln q = -kT\ln(\sum_i \exp(-\varepsilon_i/kT)), \tag{2.61}$$

which indeed is correct; the reader is referred to any standard text on statistical mechanics for a more rigorous establishment of the link between the free energy and the partition function. Our purpose here has been to use $\langle \varepsilon \rangle = u$, both of these quantities being conceptually easy to understand (being the *sum of the* average kinetic and potential energy).

Eqn (2.61) is extremely important since it expresses f (the characteristic thermodynamic function for V and T independent variables) in terms of the energy levels ε_i of quantum mechanics which themselves ultimately come from simple kinetic- and potential-energy terms in the Hamiltonian. Eqn (2.61) can also be written as

$$\exp(-f/kT) = \sum_i \exp(-\varepsilon_i/kT) = q. \tag{2.62}$$

(With regard to eqn (2.61) we recall that $\ln(abc) = \ln a + \ln b + \ln c$ but $\ln(a + b + c) \neq \ln a + \ln b + \ln c$.)

Eqns (2.58) and (2.61) give the thermodynamic parameters f and u in terms of q (which itself is given in terms of the quantum-mechanical ε_i). Using the definition $F = U - TS$ (or per molecule, $f = u - Ts$) we can solve for S and, using eqns (2.58) and (2.61), obtain S (or here s, the entropy per molecule) in terms of q (and hence in terms of the ε_i)

$$\begin{aligned} s &= (u-f)/T \\ &= -\frac{k}{T}\frac{\partial \ln q}{\partial (1/T)} + k\ln q. \end{aligned} \tag{2.63}$$

Rearrangement of eqn (2.63) using eqn (2.56) yields the basic relation for s

$$s = -k \sum_i p_i \ln p_i \qquad (2.64)$$

Since the p_i are necessarily less than unity, $\ln p_i$ is negative and $S \geqslant 1$, the equality holding at $T = 0$, where $p_1 = 1$ and all other $p_i = 0$. Eqn (2.64) can be rewritten as

$$S = k \ln \prod_i \left(\frac{1}{p_i}\right)^{p_i} = k \ln \Omega$$
$$\Omega = \prod_i \left(\frac{1}{p_i}\right)^{p_i}. \qquad (2.65)$$

If there are N states of equal probability, $p_i = 1/N$ and $S = k \ln N$ or $\Omega = N$; in general Ω is the number of accessible states of the system (at $T = 0$, where $p_1 = 1$ and all other $p_i = 0$, then $\Omega = 1$, i.e. there is only one state accessible even though there are in general an infinite number of states).

The ε_i of eqn (2.56) refer to the energy levels of an isolated atom or molecule. Thus the relations for u, f, and s we have given refer to the thermodynamic functions for such an isolated molecule and hence are not the most general relations we can obtain. The problem is that the energy levels for a system of N interacting molecules are not in general available. However, in principle we can introduce the notions of energy levels E_i with degeneracies Ω_i that are solutions of the Schrödinger equation for Avogadro's number of particles (i.e. a macroscopic sample). Then letting Q be the partition function for N particles, we have the general relations in analogy with the relations for a single molecule

$$Q = \exp(-F/kT) = \sum_i \Omega_i \exp(-E_i/kT),$$

$$U = -k \frac{\partial \ln Q}{\partial (1/T)}, \qquad (2.66)$$

$$S = -\frac{k}{T} \frac{\partial \ln Q}{\partial (1/T)} + k \ln Q.$$

Since the E_i and Ω_i are not in general available, we must consider special cases where the E_i (energy level for N molecules) can be generated in terms of the ε_i (energy levels for a single molecule) or equivalently where Q (the partition function for N molecules) can be generated in terms of q (the partition function for a single molecule). We start with a simple yet powerful model for a solid.

2.1.3. Einstein crystal

To begin our discussion of the energy levels of many-particle systems, consider the simplest (hypothetical) model possible, namely a regular crystal lattice where each particle has the same two energy levels per particle and where occupancy of either level by a given particle is independent of the states of occupancy of the levels of all the other particles (in particular the neighbours of the given particle). First consider a system composed of only two particles ($N = 2$). Then the states of occupancy for the whole (two-particle) system are

	possible states of occupancy			
particle 1	level 1	level 1	level 2	level 2
particle 2	level 1	level 2	level 1	level 2
net energy	$\varepsilon_1 + \varepsilon_1$	$\varepsilon_1 + \varepsilon_2$	$\varepsilon_2 + \varepsilon_1$	$\varepsilon_2 + \varepsilon_2$.

The partition function for the system is then

$$Q = \exp(-(\varepsilon_1+\varepsilon_1)/kT) + 2\exp(-(\varepsilon_1+\varepsilon_2)/kT) + \exp(-(\varepsilon_2+\varepsilon_2)/kT) \quad (2.67)$$

$$= \Omega_1 \exp(-E_1/kT) + \Omega_2 \exp(-E_2/kT) + \Omega_3 \exp(-E_3/kT),$$

$$\Omega_1 = 1, \quad E_1 = \varepsilon_1 + \varepsilon_1$$

$$\Omega_2 = 2, \quad E_2 = \varepsilon_1 + \varepsilon_2 \quad (2.68)$$

$$\Omega_3 = 1, \quad E_3 = \varepsilon_2 + \varepsilon_2$$

If q, the partition function for a single particle, is

$$q = \exp(-\varepsilon_1/kT) + \exp(-\varepsilon_2/kT), \qquad (2.69)$$

then we observe that Q of eqn (2.67) is generated by

$$Q = q^2. \qquad (2.70)$$

Mathematically, eqn (2.70) follows from the assumption of the independence of the particles; in general the partition function for a system of independent particles can be written at least partially as the product of partition functions for single particles. The generalization of eqn (2.70) to N particles is

$$Q = q^N. \qquad (2.71)$$

From eqn (2.71) it is clear that the thermodynamic functions for a system of N particles are the same per mole as those of a single particle for systems that can be described by eqn (2.71).

To first approximation a real crystal can be treated as a collection of particles, each particle oscillating about an equilibrium position (lattice site). Since a particle can oscillate in three directions there are $3N$ degrees of oscillatory freedom for the whole system. The potential energy for a harmonic oscillator in one dimension is given by eqn (2.52) truncated at the quadratic term

$$\varphi(x) = \tfrac{1}{2}k(x-x_0)^2. \qquad (2.72)$$

(The k in eqn (2.72) is the spring constant and is not to be confused with the Boltzmann constant.) The quantum-mechanical energy levels for this system are illustrated in Fig. 1.2(a) (p. 5) (evenly spaced levels), and the partition function has been already given in eqn (1.15). Then Q for this model of a crystal (first proposed by Einstein to explain the heat capacity of solids) is

$$Q = q^{3N} = \left(\frac{1}{1-\exp(-\Delta\varepsilon/kT)}\right)^{3N} \left(\exp(-\varphi_0/kT)\right)^N, \qquad (2.73)$$

$\Delta\varepsilon$ being the spacing between the levels. The term φ_0 is the potential energy per particle at 0 K resulting from the interaction of a particle with its neighbours.

Since eqn (2.73) gives a good first approximation to the behaviour of solids as a function of temperature it is instructive to examine the thermodynamic functions in more detail. First we introduce the parameter

$$\vartheta_V = \text{(characteristic vibrational temperature)} \quad (2.74)$$
$$= \Delta\varepsilon/k.$$

Then using eqn (2.73) in eqn (2.66) we find

$$F = N\varphi_0 + 3NkT\ln\{1-\exp(-\vartheta/T)\},$$
$$U = N\varphi_0 + 3NkT\{(\vartheta/T)/(\exp(\vartheta/T)-1)\},$$
$$S = 3Nk\{\frac{\vartheta/T}{\exp(\vartheta/T)-1} - \ln(1-\exp(-\vartheta/T))\}, \quad (2.75)$$
$$C_V = 3Nk\left(\frac{\vartheta}{T}\right)^2 \frac{\exp(\vartheta/T)}{(\exp(\vartheta/T)-1)^2}.$$

For the simple Einstein model, the crystal has a constant volume and hence $G = F$ and $H = U$ for this model (a strictly harmonic potential will give a zero coefficient of expansion in any treatment; asymmetry or anharmonic terms in the potential are required to obtain expansion of the crystal with temperature). The limits of C_V at the extremes of temperature are

$T(K)$	C_V
0	0
∞	$3Nk$

The value $3Nk$ (or $3R$ per mole) at high temperature is an experimental result known as the law of Dulong and Petit. Historically the Einstein model is very important since it explained the fact that $C_V \to 0$ as $T \to 0$, an experimental fact not predicted by classical theory (which gave $C_V = 3R$ independent of temperature); this behaviour is the simplest experimental demonstration of the need for quantization to explain the behaviour of a molecular system. In section 2.3 we will compare the results of eqn (2.75) with experimental data for real crystals with particular reference to molecular forces. Having considered a simple model for dense matter at low temperature (crystal) we now examine the extreme of low density at high temperature (gases).

2.1.4. Ideal gas

The quantum-mechanical model for a dilute gas is a point particle moving in a box with rigid walls. The energy levels for translation in one dimension are

$$\varepsilon_n = n^2 h^2 / 8mL^2 \qquad (n=1,\infty), \tag{2.76}$$

where h is Planck's constant, m the mass of the particle, and L the length of the box. The partition function for the translation of a single particle in one dimension is

$$q_{tr} = \sum_{n=1}^{\infty} \exp(-n^2 h^2 / 8mL^2 kT). \tag{2.77}$$

For a molecule in a macroscopic box, $\varepsilon_n/kT \ll 1$, hence the sum in eqn (2.77) can be accurately evaluated by replacing the sum by an integral. We readily obtain

$$q_{tr} = \left(\frac{2\pi mkT}{h^2}\right)^{\frac{1}{2}} \frac{1}{L}. \tag{2.78}$$

Eqn (2.78) gives the partition function for a single molecule translating in one dimension; for a single molecule translating in three dimensions, where translation in each dimension is independent of the others, we have

$$q_{tr} = \left(\frac{2\pi mkT}{h^2}\right)^{\frac{3}{2}} \frac{1}{L^3}. \tag{2.79}$$

Taking

$$V = L^3,$$
$$\Lambda = \left(\frac{2\pi mkT}{h^2}\right)^{-\frac{1}{2}}, \tag{2.80}$$

we have

$$q_{tr} = V/\Lambda^3 = Vq,$$
$$q = 1/\Lambda^3.$$
(2.81)

Note that partition functions are always dimensionless numbers; hence Λ^3 has the dimensions of volume, Λ having the dimensions of length (this quantity is referred to as the de Broglie thermal wavelength).

The quantity q_{tr} in eqn (2.81) is the partition function (sum over translational energy levels) for a single molecule in a box of volume V. The partition function for N independent particles in a box is given by

$$Q = (Vq)^N/N! \ .$$
(2.82)

The factor $(Vq)^N$ arises from the assumption that the particles are independent, i.e. we can neglect all interactions between them, for the same reason as explained in obtaining eqn (2.71). But eqn (2.82) contains an important difference from eqn (2.71), namely the division by $N!$. This factor is present in the Q for the ideal gas and not in the Q for the Einstein crystal for the following reason. In the gas model each particle can be in any translational state; for every particular assignment of states to the whole system (particle a in state i, b in state j, etc.) the product $(Vq)^N$ generates all possible combinations of swapping particle labels (particle b in state i, a in state j, etc.), there being exactly $N!$ ways to swap particle labels among N particular states. Since the particles are in reality identical, the swapping of labels does not produce a new state of the total system. Thus $(Vq)^N$ over-counts allowed states by exactly the factor $N!$, the over-counting being corrected for by division by $N!$. This over-counting does not arise in the crystal model, since the swapping of labels does produce a new state of the system, each particle having a particular location in space (lattice site) in the crystal.

From eqns (2.80)–(2.82) we obtain

$$Q = \exp(-F/kT) = \left(\frac{2\pi mkT}{h^2}\right)^{3N/2} \frac{V^N}{N!}. \qquad (2.88)$$

Using Stirling's approximation for $N!$

$$N! \simeq N^N e^{-N},$$
$$\ln N! \simeq N\ln N - N, \qquad (2.84)$$

which for N of the order of 10^{24} is an extraordinarily good approximation, we find

$$Q = \exp(-F/kT) = \left\{\left(\frac{2\pi k}{h^2}\right)^{3/2} e\left(\frac{m^{3/2}T^{3/2}}{\rho}\right)\right\}^N, \qquad (2.85)$$

where $\rho = N/V$ is the density.

From eqn (2.85) and eqns (2.21) and (2.25) we obtain all the well-known relations for the ideal gas:

$$F = NkT\ln\Lambda^{3/2} - NkT + NkT\ln\rho,$$
$$U = \tfrac{3}{2}NkT,$$
$$S = -Nk\ln\Lambda^{3/2} + \tfrac{5}{2}Nk - Nk\ln\rho, \qquad (2.86)$$
$$\mu = kT\ln\Lambda^{3/2} + kT\ln\rho,$$
$$pV = NkT \text{ (or } p/\rho kT = 1\text{)},$$
$$C_V = \tfrac{3}{2}Nk.$$

Of particular importance to our discussion of equilibrium is the form of the chemical potential in eqn (2.86) which is seen to be identical with that quoted in eqn (2.41). The following are concomitant:

N-dependence of the partition function	$Q = (Vq)^N/N!,$	
density-dependence of the chemical potential	$\mu = \mu^0(T) + kT\ln\rho,$	(2.87)
relation between densities at equilibrium (law of mass action)	$\prod_i \rho_i^{\nu_i} = K(T).$	

When we talk of ideal gases or solutions, all the relations in eqn (2.87) apply (they are, of course, related by the general relations $F = -kT\ln Q$, $\mu = \partial F/\partial N$, $\Sigma \mu_i \nu_i = 0$).

Evaluating the constants in the equation for S in eqn (2.86) we have (per mole)

$$S = R\{1.343 + \tfrac{3}{2}\ln T + \tfrac{3}{2}\ln(\text{molecular weight}) - \ln(\text{concentration})\}, \tag{2.88}$$

where the concentration is given in the practical units of mol l^{-1}. In section 2.2 numerical examples of the thermodynamic functions for the ideal gas will be given, in particular the entropy. Eqn (2.88) is known as the Sackur–Tetrode equation.

2.1.5. A gas-to-crystal model; maximum-term approximation

We have examined two simple models which have total-system partition functions

$$\begin{cases} \text{low } T \\ \text{high } \rho \end{cases} \text{(crystal)} \quad Q = q^N \text{ (harmonic oscillator } q\text{)},$$

$$\begin{cases} \text{high } T \\ \text{low } \rho \end{cases} \text{(gas)} \quad Q = (Vq)^N/N! \text{ (particle in a box } q\text{)}. \tag{2.89}$$

We now give a very simple model which in the limits of high and low density respectively gives the results of eqn (2.89). The model is that of a regular lattice of sites where a lattice site can either be occupied or not occupied by a particle. We further assume that a single particle on a lattice site is described by a partition function q that is independent of the state of occupancy of neighbouring sites. If there are M lattice sites, then the exact partition function for N particles on M sites is given by

$$Q = \frac{M!}{(M-N)!N!} q^N, \tag{2.90}$$

where the combinatorial factor gives the number of distinct ways N particles can be placed on M sites. Let the fraction of sites occupied be

$$\vartheta = N/M, \tag{2.91}$$

which can vary from zero to unity and plays the role of the density in this model. Using the definition of ϑ in eqn (2.91) and Stirling's approximation for the factorials (eqn (2.84)), we obtain

$$Q = \left[\left(\frac{1}{1-\vartheta}\right)^{1/\vartheta}\left(\frac{1-\vartheta}{\vartheta}\right)q\right]^N. \qquad (2.92)$$

In the limits of $\vartheta \to 0$ and $\vartheta \to 1$ we have

$$\lim_{\vartheta \to 1} Q = q^N, \qquad (2.93)$$

$$\lim_{\vartheta \to 0} Q = (eq/\vartheta)^N = (Mq)^N/N!,$$

where in the lim $\vartheta \to 0$ we have used the fact that $(1/1-\vartheta)^{1/\vartheta}$ approaches e as $\vartheta \to 0$ and that $(e/\vartheta)^N = (eM/N)^N$, $(e/N)^N$ being Stirling's approximation for $1/N!$. Thus eqn (2.90), or equivalently eqn (2.92), is the general partition function for this particular model valid at all $\vartheta = 0, 1$. In the limits of $\vartheta \to 0$ and $\vartheta \to 1$ this partition function takes on the asymptotic forms for a crystal ($\vartheta = 1$) and dilute gas ($\vartheta \to 0$) as given in eqn (2.89). In general we cannot write down an exact partition function for a real system valid at all densities. This model, however, serves to show that eqn (2.89) can be obtained as opposite limits of the same partition function and should be thought of as limiting forms for high and low density and not necessarily qualitatively different entities.

The free energy for the system described by eqn (2.92) is

$$F = -kT\ln Q = -NkT\ln q - MkT\{-\vartheta\ln\vartheta - (1-\vartheta)\ln(1-\vartheta)\}. \qquad (2.94)$$

The quantity

$$S_{mix} = -Mk\{\vartheta\ln\vartheta + (1-\vartheta)\ln(1-\vartheta)\} \qquad (2.95)$$

is known as the entropy of mixing for an ideal system (independent particles) and is a simple example of eqn (2.64). The entropy of mixing takes on the values

$$S_{mix} = \begin{cases} 0, & \text{for } \vartheta = 0 \\ Mk\ln 2, & \vartheta = \frac{1}{2} \\ 0, & \text{for } \vartheta = 1 \end{cases} \quad (2.96)$$

S_{mix} is always positive (since ϑ and $1 - \vartheta$ are less than unity ln ϑ and ln$(1-\vartheta)$ are always negative) and has its maximum value at $\vartheta = \frac{1}{2}$; for the limits at $\vartheta = 0$ and $\vartheta = 1$ we have used the fact that as x goes to zero, $x \ln x = 0$.

We will encounter the above lattice model again since it serves as a primitive model for binding to membranes (two-dimensional lattice) or polymers (one-dimensional lattice). Returning to our very simple model of a crystal (independent particles on lattice sites with two energy levels per particle) we can use a modification of eqn (2.90) to illustrate the relation between the sum over states for a system of N particles and the sum over states for a single particle. Let there be N lattice sites (which is also the total number of particles) with N_1 and N_2 the number of particles in states one and two respectively. Unlike the model described by eqn (2.90), where the number of different lattice sites (in that model, occupied or unoccupied) was fixed, we must allow N_2 to take on all values from $N_2 = 0$ to $N_2 = N$ (with $N_1 + N_2 = N$). Thus we have

$$Q = \sum_{N_2=0}^{N} \frac{N!}{(N-N_2)!N_2!} \exp(-(N-N_2)\varepsilon_2/kT) \exp(-N_2\varepsilon_2/kT)$$

$$= (\exp(-\varepsilon_1/kT))^N \sum_{N_2=0}^{N} \frac{N!}{(N-N_2)!N_2!} x^{N_2} \quad (2.97)$$

($x = \exp(-\Delta\varepsilon/kT)$, $\Delta\varepsilon = \varepsilon_2 - \varepsilon_1$).

The sum in eqn (2.97) is the binomial expansion (eqn 1.83)) and hence we can obtain a closed expression for Q

$$Q = (\exp(-\varepsilon_1/kT))^N (1+x)^N$$
$$= \{(\exp(-\varepsilon_1/kT) + \exp(-\varepsilon_2/kT)\}^N = q^N, \quad (2.98)$$

which, of course, reproduces eqn (2.71).

Eqn (2.97) gives a specific example of Q in eqn (2.66) with the identification

$$\Omega(N,N_2) = \frac{N!}{(N-N_2)!N_2!},$$
$$E(N_2) = (N-N_2)\epsilon_1 + N_2\epsilon_2. \qquad (2.99)$$

Eqn (2.97) can be written as

$$Q = \sum_{N_2=0}^{N} t(N,N_2), \qquad (2.100)$$

$$t(N,N_2) = \Omega(N,N_2)\exp(-E(N_2)/kT).$$

To find what value of N_2 maximizes $t(N,N_2)$ we set

$$\partial \ln t(N,N_2)/\partial N_2 = 0. \qquad (2.101)$$

Using Stirling's approximation for the factorials, eqn (2.101) gives

$$\frac{N_2^*}{N} = \frac{\exp(-\epsilon_2/kT)}{\exp(-\epsilon_1/kT) + \exp(-\epsilon_2/kT)}, \qquad (2.102)$$

where N_2^* is the value of N_2 that maximizes $t(N,N_2)$. Comparing this with the exact value of $\langle N_2 \rangle$

$$\langle N_2 \rangle = \frac{\Sigma N_2 t(N,N_2)}{\Sigma t(N,N_2)} \qquad (2.103)$$

$$= \frac{\partial \ln Q}{\partial \ln x} = N \frac{\exp(-\epsilon_2/kT)}{\exp(-\epsilon_1/kT) + \exp(-\epsilon_2/kT)},$$

we see that

$$\langle N_2 \rangle = N_2^*, \qquad (2.104)$$

which states that the average value of N_2 is the same as the most probable value (the value that maximizes $t(N,N_2)$). Further, if we define

$$\vartheta^* = N_2^*/N \qquad (2.105)$$

then the single term of Q for which $N_2 = N_2^*$ is (using Stirling's approximation for the factorials)

$$\frac{N!}{(N-N_2^*)!N_2^*!}\left(\exp\{-(\epsilon_2-\epsilon_1)/kT\}\right)^{N_2^*}\left(\exp\{-\epsilon_1/kT\}\right)^N$$

$$= \left\{\left(\frac{1}{1-\vartheta^*}\right)\left(\frac{1-\vartheta^*}{\vartheta^*}\right)^{\vartheta^*}\left(\exp\{-\epsilon_2/kT\}\right)^{\vartheta^*}\left(\exp\{-\epsilon_1/kT\}\right)^{1-\vartheta^*}\right\}^N \qquad (2.106)$$

$$= \left\{\exp(-\epsilon_1/kT) + \exp(-\epsilon_2/kT)\right\}^N = q^N ,$$

which is a remarkable result in that it states that one term of the sum over all N_2 is equal to the total sum.

This property of being able to replace a partition function by its maximum term is a general one. We must note very carefully, however, that this is only true for the partition function for a total system of N particles ($N \sim O(10^{24})$) and is not true for the partition function for a single particle (i.e. in our notation it is true for Q but not for q). We can easily see that when we are talking about the occupation probability distribution for a single molecule, the distribution is quite broad (see Figs 1.3 and 1.5, pp. 9 and 13), and it would be very incorrect to keep only the most probable state (e.g. in the harmonic oscillator model in Fig. 1.3 this would always be the ground state). A general statement[5] of the above result is that (N_i a summation index for the number of particles in a general state i)

$$\left(\frac{1}{N}\right)kT\ln Q(N,N_i) = \left(\frac{1}{N}\right)kT\ln t(N,N_i^*) + O\left(\frac{1}{\sqrt{N}}\right), \qquad (2.107)$$

i.e. the free energy per particle is given accurately in terms of the maximum term of Q with an error of the order of $1/\sqrt{N}$; since $N \sim O(10^{24})$, $1/\sqrt{N} \sim O(10^{-12})$, which is an error beyond the limits of experimental accuracy. Incidentally in our example we found that the maximum term (eqn (2.106)) gave Q (eqn (2.98)) exactly; this is a result of having used Stirling's approximation which in this example fortuitously absorbed the $1/\sqrt{N}$ error. Eqn (2.107) is a general result of probability theory for N large and is known as the central limit theorem. The result of eqn (2.107) is of enormous practical importance since it means that when we are writing the partition function for a total system we need not obtain Q in closed form, the maximum term being sufficient to give results that are extraordinarily accurate. However, in biophysical examples it is most often the quantity q, the molecular partition function, that is of interest, and in using this quantity we must use the total partition function.

2.1.6. Grand partition function

There is another partition function that is much used in current research in statistical mechanics, but which is not very familiar to non-specialists. In fact simple forms of this partition function have already been extensively used in Chapter I without introducing it by name. Here we want to introduce it formally since it is very useful in applications. The partition function we are discussing is called the grand partition function and is given by[6]

$$\Xi(V,T) = \sum_N Q(N,V,T)\lambda^N \qquad (2.108)$$

where

$$\Xi = \exp(pV/kT),$$

$$Q = \exp(-F/kT), \qquad (2.109)$$

$$\lambda = \exp(\mu/kT).$$

The quantity λ is sometimes referred to as the absolute activity. The grand partition function is equally useful in treating binding (adsorption or titration) problems where the number of particles bound is a variable or where N is fixed such as in a sample of gas with N particles in a box of volume V in equilibrium with a bath of temperature T. Since Ξ (eqn (2.109)) involves a sum over N, the application to systems of fixed N seems contradictory. As with eqn (2.107) for Q, if Ξ is the grand partition function for a macroscopic system then

$$\left(\frac{1}{N^*}\right)\ln\Xi = \left(\frac{1}{N^*}\right)\ln Q(N^*) + \ln\lambda + O\left(\frac{1}{\sqrt{N}}\right), \qquad (2.110)$$

i.e. while Ξ is formally a sum over N, only the term in Ξ that maximizes $Q\lambda^N$ contributes significantly, the error again being of the order of $1/\sqrt{N}$ (thus if $N^* \sim 10^{24}$ the error is $\sim 10^{-12}$ or negligible). The usefulness of the grand partition function lies in our being able to force $Q\lambda^N$ to have its maximum value at any value of N^* we wish. This is achieved using the standard method of calculating averages in statistical mechanics

$$\langle N \rangle = \frac{\Sigma N Q \lambda^N}{\Sigma Q \lambda^N} = \frac{\partial \ln \Xi}{\partial \ln \lambda} \qquad (2.111)$$

or

$$N^* = \frac{\partial \ln \Xi}{\partial \ln \lambda}, \qquad (2.112)$$

which if Ξ is being used to treat a system of fixed, predetermined N^*, is an equation for λ. From eqn (2.109) note that

$$\left(\frac{1}{N^*}\right)\ln\Xi = \frac{pV}{N^* kT} \qquad (2.113)$$

gives the equation of state.

To illustrate the use of the grand partition function let us treat the ideal gas. From eqn (2.82), which gives the N dependence of Q for N particles in a box, we have

$$\Xi = \sum_{N=0}^{\infty} \frac{(Vq)^N}{N!} \lambda^N = \sum_{N=0}^{\infty} \frac{(Vq\lambda)^N}{N!}. \qquad (2.114)$$

Note that at this point λ is unknown while we know (or have chosen arbitrarily) the value of N we want to discuss. Since

$$e^x = \sum_{n=0}^{\infty} \frac{x^n}{n!}, \tag{2.115}$$

eqn (2.114) can be evaluated to give

$$\Xi = \exp(Vq\lambda) = \exp(pV/kT) \tag{2.116}$$

or

$$\ln\Xi = Vq\lambda. \tag{2.117}$$

Applying eqn (2.113),

$$\frac{\partial \ln\Xi}{\partial \ln\lambda} = Vq\lambda = N^* \tag{2.118}$$

or

$$\lambda = \left(\frac{N^*}{V}\right)\frac{1}{q}, \tag{2.119}$$

which gives the unknown λ in terms of the known quantities N^*, V, and q. Since $\lambda = \exp(\mu/kT)$ we have immediately from eqn (2.119)

$$\mu = -kT\ln q + kT\ln\rho, \tag{2.120}$$

where $\rho = N^*/V$. Eqn (2.118) in eqn (2.113), of course, yields the ideal gas law; note that the derivation of $pV = NkT$ did not require any knowledge of q, only the N dependence of Q (eqn (2.82)), which follows in general from the assumption of independent particles (i.e. $pV = NkT$ follows solely from the assumption of independence and not from any special properties of the particle in a box).

The thermodynamic potentials of the various partition functions can be easily remembered using the maximum-term approximation (eqns (2.107) and (2.110))

$$\begin{aligned}\ln Q &\simeq \ln\Omega(N^*) - E(N^*)/kT \\ (-F/kT &= S/k - U/kT), \\ \ln\Xi &\simeq \ln Q(N^*) + N^*\ln\lambda \\ (pV/kT &= -F/kT + G/kT),\end{aligned} \tag{2.121}$$

AND INTERMOLECULAR FORCES 91

which are just the definitions of F and G, $F = U-TS$ and $G = F + pV$.

2.1.7. Physical cluster model from the grand partition function

Since the physical cluster model introduced in Chapter 1 is applicable to many systems of interest in biophysical chemistry, at this point we want to give a more fundamental development. To begin we first note that the partition function for a mixture of ideal gases having a fixed number of molecules of each species (no reactions taking place) is given as the product of the partition functions for the individual species

$$Q(N_1,N_2,\ldots) = \frac{(Vq_1)^{N_1}}{N_1!} \frac{(Vq_2)^{N_2}}{N_2!} \ldots = \prod_i \frac{(Vq_i)^{N_i}}{N_i!}. \quad (2.122)$$

Now consider a single species which can self-aggregate, as in eqn (1.104) to form dimers, trimers, etc. Letting

N = total number of monomer units,
N_1 = number of free monomer units,
N_2 = number of dimer units,
$\ldots$
N_n = number of n-mer units,
$\ldots$
q_1 = partition function for a single monomer,
q_2 = partition function for a single dimer,
$\ldots$
q_n = partition function for a single n-mer,
$\ldots$

then for fixed $N_1, N_2, \ldots N_n, \ldots$,

$$t(N_1,N_2,\ldots,V,T) = \prod_n \frac{(Vq_n)^{N_n}}{N_n!}. \quad (2.123)$$

Since the molecules are free to react, we cannot restrict the

N_n to specified values. Hence the general partition function must allow all values of N_n

$$Q(N,V,T) = \sum_{N_n} \prod_n \frac{(Vq_n)^{N_n}}{N_n!} = \sum_{N_n} t(N,N_n), \qquad (2.124)$$

subject to the conservation of total monomer units

$$\sum_n nN_n = N. \qquad (2.125)$$

A closed form for the sum in eqn (2.124) cannot be found with the constraint of eqn (2.125). However, since the N_n are numbers of the order of 10^{24}, we can use the maximum-term approximation and find the set of N_n that maximizes the term in eqn (2.123). Since we must maximize $t(N_n)$ subject to the constraint of eqn (2.125), the method of Lagrange undetermined multipliers is required, which introduces the influence of the constraint into the maximization via an unknown parameter which with foresight we choose to call $\ln \lambda$

$$\frac{\partial \ln t(N,N_n)}{\partial N_n} + \ln \lambda \frac{\partial \Sigma n N_n}{\partial N_n} = 0. \qquad (2.126)$$

Again using Stirling's approximation for the factorials, eqn (2.126) becomes

$$\frac{\partial}{\partial N_n} \left\{ \sum_N (N_n \ln Vq_n - N_n \ln N_n + N_n) + \ln \lambda \Sigma n N_n \right\}$$
$$= \ln Vq_n - \ln N_n + n \ln \lambda = 0 \qquad (2.127)$$

or

$$Vq_n \lambda^n = N_n^*. \qquad (2.128)$$

To determine the unknown λ, we insert N_n^* of eqn (2.128) into eqn (2.125), giving

$$\sum_{n=1}^{\infty} nVq_n \lambda^n = N \qquad (2.129)$$

or

$$\sum_{n=1}^{\infty} n q_n \lambda^n = c_0, \qquad (2.130)$$

where $c_0 = N/V$. Eqn (2.130) represents an equation for the unknown λ in terms of the known quantities c_0 and the partition functions for clusters, the q_n. We will return to the physical meaning of eqn (2.130) shortly, but now we want to repeat the derivation using Ξ which will illustrate how simple the use of the grand partition function makes many problems.

Using eqn (2.114) with the Q of eqn (2.124), we have immediately

$$
\begin{aligned}
\text{(a)} \quad & \Xi = \sum_{N=0}^{\infty} Q \lambda^N, \\
\text{(b)} \quad & = \sum_{N=0}^{\infty} \left[\sum_{N_n} \prod_n \frac{(V q_n)^{N_n}}{N_n!} \right] \lambda^N \\
\text{(c)} \quad & = \sum_{N=0}^{\infty} \sum_{N_n} \prod_n \frac{(V q_n \lambda^n)^{N_n}}{N_n!} \\
\text{(d)} \quad & = \prod_n \sum_{N_n=0}^{\infty} \frac{(V q_n \lambda^n)^{N_n}}{N_n!} \\
\text{(e)} \quad & = \prod_n \exp(V q_n \lambda^n) \\
\text{(f)} \quad & = \exp\left(\sum_n V q_n \lambda^n\right) \\
\text{(g)} \quad & = \exp(pV/kT)
\end{aligned}
\qquad (2.131)
$$

or

$$pV/kT = V \sum_n q_n \lambda^n, \qquad (2.132)$$

eqn (2.132) giving a general equation of state for a gas whose units are associated to give independent clusters. The steps taken in eqn (2.131) are: (a) general definition of Ξ (eqn (2.108)); (b) substitution of the specific form of Q (eqn (2.124)) for this problem; (c) distribution of the factors λ^N using eqn (2.125); (d) interchanging the operations of product and sum using the fact that the sum over N and the sum over the N_n generate all possible values of the N_n; (e) closure of the sums over N_n utilizing eqn (2.115); (f) use of the fact that the product of exponentials is the exponential of the sum of exponents; and (g) general thermodynamic potential of the grand partition function (eqn (2.109)). The important property of the grand partition function that makes it useful is that, while we could not obtain a closed form for Q (eqn (2.124)) subject to the constraint of eqn (2.125), on dropping the constraint and summing over all possible values of the N_n, Ξ is readily evaluated. The influence of the constraint is introduced by using the value of λ that satisfies eqn (2.112). Using eqn (2.112) we have

$$N^* = \sum_n nVq_n \lambda^n \tag{2.133}$$

or

$$\sum_n nq_n \lambda^n = N^*/V = c_0, \tag{2.134}$$

which is identical with eqn (2.130); thus the undetermined multiplier $\ln\lambda$ is μ/kT (in general undetermined multipliers have physical significance).

Note that eqn (2.132) can be interpreted as

$$p/kT = \sum_n q_n \lambda^n = \Sigma\rho_n, \tag{2.135}$$

where ρ_n is the concentration or density of clusters containing n molecules ($\rho_n = N_n^*/V$); eqn (2.135) is simply Dalton's law of partial pressures, obtained here since we assumed the clusters are independent. Thus we have

$$\rho_n = q_n \lambda^n, \tag{2.136}$$

or in particular for $n = 1$

$$\rho_1 = q_1 \lambda, \quad (2.137)$$

which gives

$$\lambda = \rho_1/q_1, \quad (2.138)$$

or, since $\lambda = \exp(\mu/kT)$,

$$\mu = -kT\ln q_1 + kT\ln \rho_1. \quad (2.139)$$

Notice that μ is the chemical potential for the system as a whole

$$\mu = kT \frac{\partial \ln Q}{\partial \ln N^*} \quad (2.140)$$

and that in this particular model (clusters assumed independent) the chemical potential for N^* monomer units in any form is the same as the chemical potential for the free monomer. From eqn (2.126), using the fact that $kT\partial \ln t(N_n)/\partial N_n$ is the chemical potential for an n-mer, we obtain

$$\mu_n = kT\ln \lambda^n = -kT\ln q_n + kT\ln \rho_n. \quad (2.141)$$

If we write the assocation reaction for the n-mer (eqn (1.104))

$$nA \rightleftharpoons A_n, \quad (2.142)$$

then eqn (2.32) demands

$$\mu_n = n\mu_1$$
$$(kT\ln \lambda^n = nkT\ln \lambda), \quad (2.143)$$

which is seen to be consistent with the form of μ_n given in eqn (2.141).

Using the form of λ given in eqn (2.138) in eqn (2.134), or equivalently eqn (2.130), we have

$$\sum_{n=1}^{\infty} n[A_1]^n \left(\frac{q_n}{q_1^n}\right) = c_0. \tag{2.144}$$

Defining

$$K_n = \frac{q_n}{q_1^n} \tag{2.145}$$

gives

$$\sum_{n=1}^{\infty} n[A_1]^n K_n = c_0, \tag{2.146}$$

which is eqn (1.108), derived from considering multiple equilibria (eqn (1.104)).

2.1.8. Molecule grand partition function Ξ

In Chapter 1 we extensively used a partition function denoted as Ξ to treat the dissociation of polyprotic acids. We now derive this partition function from fundamental considerations and establish its relation to Ξ. Consider molecules A which can bind any number of molecules B. For the species AB_n (A with n Bs bound) let

N_n = number of A molecules with n Bs bound

q_B = partition function for a free B molecule

q_n = partition function for a single AB_n complex

N_B = number of free (unbound) B molecules.

Then Ξ for this system is

$$\Xi = \sum_{N_B} \frac{(Vq_B)^{N_B}}{N_B!} \lambda_B^{N_B} \sum_{N_n} \prod_n \frac{(Vq_n)^{N_n}}{N_n!} \lambda_A^{N_n} \lambda_B^{nN_B}. \tag{2.147}$$

Eqn (2.147) represents a summation over all possible numbers

AND INTERMOLECULAR FORCES   97

N_B and N_n of the species B and AB_n; note that for every A and B molecule there is respectively a factor λ_A and λ_B assigned. In analogy with eqn (2.112) the factors λ_A and λ_B are determined from the relations

$$N_A^* = \frac{\partial \ln \Xi}{\partial \ln \lambda_A},$$
$$N_B^* = \frac{\partial \ln \Xi}{\partial \ln \lambda_B},$$
(2.148)

where N_B^* and N_A^* are the total numbers of A and B respectively in any form (known variables under experimental control). Following the same steps used in eqn (2.131) we readily find

$$\Xi = \exp(Vq_B \lambda_B) \exp(\sum_n Vq_n \lambda_B^n \lambda_A).$$
(2.149)

Using eqn (2.148) we have

$$c_B = q_B \lambda_B + \lambda_A \sum_n nq_n \lambda_B^n,$$
$$c_A = \lambda_A \sum_n q_n \lambda_B^n,$$
(2.150)

where $c_A = N_A^*/V$, $c_B = N_B^*/V$. Using the second part of eqn (2.150) to eliminate λ_A from the first we obtain

$$c_B = q_B \lambda_B + c_A \frac{\Sigma nq_n \lambda_B^n}{\Sigma q_n \lambda_B^n},$$
(2.151)

which is an implicit equation for λ_B in terms of the knowns c_A, c_B, q_B, and the q_n. Solving eqn (2.151) for λ_B, λ_A is given by (using eqn (2.150))

$$\lambda_A = c_A / \Sigma q_n \lambda_B^n.$$
(2.152)

From the obvious physical interpretation of eqn (2.150) as

$$c_B = [B] + \Sigma n[AB_n],$$
$$c_A = \Sigma [AB_n],$$
(2.153)

we have

$$\lambda_B = [B]/q_B,$$

$$\lambda_A \lambda_B^n = [AB_n]/q_n. \qquad (2.154)$$

The probability of AB_n is then

$$p[AB_n] = \frac{[AB_n]}{\Sigma[AB_n]} = \frac{\lambda_A \lambda_B^n q_n}{\lambda_A \Sigma q_n \lambda_B^n}$$

$$= \frac{\lambda_B^n q_n}{\Sigma q_n \lambda_B^n} = \frac{[B]^n q_n/q_B^n}{\Sigma[B]^n q_n/q_B^n}. \qquad (2.155)$$

Notice that the cluster probability distribution does not depend explicitly on λ_A, though λ_B is a function of c_A through eqn (2.151). For the reaction

$$A + nB \rightleftharpoons AB_n, \qquad K_n = q_n/q_B^n, \qquad (2.156)$$

defining

$$\Xi = \sum_n q_n \lambda_B^n = \Sigma[B]^n K_n, \qquad (2.157)$$

then the average extent of binding is given as

$$\langle n \rangle = \frac{\Sigma n q_n \lambda_B^n}{\Sigma q_n \lambda_B^n} = \frac{\partial \ln \Xi}{\partial \ln \lambda_B} = \frac{\partial \ln \Xi}{\partial \ln [B]}. \qquad (2.158)$$

The quantity Ξ plays the role of the grand partition function per molecule, eqn (2.157) representing a sum over all states of binding for a single molecule (compare eqns (2.157) and (1.62)). For binding of protons, $B = H^+$, and $[H^+]$ is experimentally available simply by measuring the pH. In general the concentration of free B is not a quantity that can be experimentally measured, and eqn (2.151) must be used to calculate λ_B, $[B] = q_B \lambda_B$.

The derivations we have just given illustrate how multiple binding and self-association can be treated from the point of view of writing the grand partition function for the total system. For systems where the clusters or aggregates are assumed independent we obtain the same results (as was done in Chapter 1) simply by systematically considering the equilibrium constants for all possible degrees of aggregation. It is important, however, to realize how the results come from a total system partition function, since if the assumption of independence is dropped, the grand partition function is the only systematic way to treat such systems (recall that the law of mass action — equilibrium constants — applies only to independent molecules or clusters). Thus

$$\Xi = \sum_N Q_N \lambda^N$$

is a general relation with λ given implicitly by

$$N^* = \partial \ln \Xi / \partial \ln \lambda,$$

the Q_N being given in general terms of the quantum-mechanical energy levels of the system. Only for independent systems is Ξ (or ξ) (eqn (2.157)) given as a power series in the concentration, that is eqn (2.154) and (2.136) are not general, applying only to systems with independent components.

2.1.9. Classical statistical mechanics

The partition functions we have been discussing are sums over quantum-mechanical energy levels as illustrated by eqn (2.56) for q (with Q then given in terms of q, Ξ in terms of Q). For vibration in crystals and molecules it is essential to use quantum mechanics. However, for liquids and solutions the effects of quantization on the motion of molecules interacting through weak intermolecular forces (not chemical bonds) is very small and these systems can accurately be described by classical mechanics. For a monatomic fluid of N particles, the formation of Q begins with the formulation of the Hamiltonian (eqn (2.53)) for the classical translation of and interaction between the atoms (not including terms for the interaction of the electrons and protons within the atom, this energy being essentially constant)

$$\mathcal{H}(p_i, r_i) = \sum_i^{3N} p_i^2/2m + \Phi(r_i), \qquad (2.159)$$

where the summation indicates that each particle has three components of momentum, hence a total of $3N$ degrees of translational freedom. The symbol $\Phi(r_i)$ represents the total potential energy of the system and is a function of the position vectors r_i of all the particles. The classical partition function is given by (the integration being over all the p_i and r_i)

$$Q = \frac{1}{h^{3N}N!} \int_{p_1} \ldots \int_{r_N} \exp(-\mathcal{H}(p_i, r_i)/kT)\, dp_1 \ldots dr_N. \qquad (2.160)$$

Since the p_i and r_i dependence of $\mathcal{H}$ are separate, the integrals over the p_i and r_i can be independently evaluated. The integral over the $3N$ p_i gives

$$\frac{1}{h^{3N}} \left(\int_{-\infty}^{\infty} \exp(-p^2/2mkT)\, dp \right)^{3N} = \left(\frac{2\pi mkT}{h^2} \right)^{3N/2}, \qquad (2.161)$$

giving Q as

$$Q = Z/\Lambda^{3N}, \qquad (2.162)$$

where Λ is given by eqn (2.80) and Z is

$$Z = \frac{1}{N!} \int_{r_1} \ldots \int_{r_N} \exp(-\Phi(r_i)/kT)\, dr_1 \ldots dr_N. \qquad (2.163)$$

The quantity Z is known as the configuration integral. Physically, Z involves the integration of each particle over the whole volume available, each configuration weighted using $\Phi(r_i)$, the potential energy for the entire system; the factor $1/N!$ arises for the same reason (correcting for label-swapping) as discussed in conjunction with eqn (2.82).

For an ideal gas, where we assume that there are no interactions between the particles, $\Phi(r_i) = 0$ and

$$Z = V^N/N!,$$
$$Q = V^N/\Lambda^{3N}N!, \qquad (2.164)$$

giving the same Q obtained from the quantum-mechanical energy levels for a particle in a box (eqn (2.82)). The integration of each particle over V with $\exp(-\Phi/kT) = 1$ gives V for each particle and V^N for N particles.

$$\mathcal{H}(p_i, \mathbf{r}_i) = \mathcal{H}(p_i) + \mathcal{H}(\mathbf{r}_i), \qquad (2.165)$$

which states that the p_i and $\mathbf{r}_i$ dependence are separate, implies that the free energy similarly can be written as a sum of contributions from the momentum terms (kinetic energy) and configuration terms (potential energy). Specifically

$$F = -kT\ln Q = 3NkT\ln\Lambda - kT\ln Z. \qquad (2.166)$$

Similarly, since all the other thermodynamic functions are given as appropriate derivatives of F (see eqn (2.21)), all the thermodynamic functions for a simple monotonic fluid separate into contributions from kinetic and potential energy. In particular the internal energy is

$$U = \tfrac{3}{2}NkT + \langle \Phi \rangle, \qquad (2.167)$$

where $\tfrac{3}{2}NkT$ is the average kinetic energy and $\langle \Phi \rangle$ is the average potential energy. The average kinetic energy is $\tfrac{3}{2}NkT$ independent of the density (gas or liquid); of course the average potential energy depends on the density, since the density determines the average distance between particles.

Note that any classical degree of freedom of the form $\varepsilon = ax^2$ contributes $(\tfrac{1}{2}kT)$ to the internal energy:

$$\langle \varepsilon \rangle = \frac{\int_0^\infty ax^2 \exp(-ax^2/kT)\,dx}{\int_0^\infty \exp(-ax^2/kT)\,dx} = \tfrac{1}{2}kT. \qquad (2.168)$$

The result of eqn (2.168) is known as the equipartition of energy and is an important result of classical statistical mechanics. Since there are $3N$ degrees of translational freedom for N particles, each degree of freedom contributes $\frac{1}{2}kT$ to U, the total contribution being $\frac{3}{2}NkT$, as indicated in eqn (2.167).

The classical formulation of statistical mechanics is easier to visualize since the quantity of interest (eqn (2.163)) involves the integration over all the configurations of N particles in a volume V, each configuration being weighted using the appropriate potential energy. It also simplifies the discussion of solutions, a topic we now turn to.

2.1.10. Dilute solutions

While we have been stressing the usefulness of the grand partition function, all the examples we have given are for aggregation or binding in the gas phase. We now want to show that the same formulae are applicable for molecules in dilute solution. Formally we can write the configuration integral for a solution of two species as

$$Z = \frac{1}{N_1!} \frac{1}{N_2!} \int_{\mathbf{r}_1} \int_{\mathbf{r}_2} \exp(-\Phi(\mathbf{r}_1,\mathbf{r}_2)/kT) d\mathbf{r}_1 \, d\mathbf{r}_2 , \qquad (2.169)$$

where the subscripts in eqn (2.169) refer to species type, there being N_1 and N_2 integrals respectively over the symbolic $\mathbf{r}_1$ and $\mathbf{r}_2$. If we take $N_1 \gg N_2$ (thus N_1 plays the role of solvent, N_2 the role of solute) we can neglect interactions between particles of species 2. Then eqn (2.169) can be written as

$$Z = \exp(-F_{conf}(N_1)/kT) \frac{(Vq)^{N_2}}{N_2!} , \qquad (2.170)$$

where $F_{conf}(N_1)$ is the configurational free energy of the pure solvent and q is

$$q = \exp(-\Delta F/kT), \qquad (2.171)$$

where ΔF represents the free energy arising from the intrinsic energy levels of species 2 and from the mutual effect of the solvent and solute on each other. Thus formally the partition for a dilute solution has the same N dependence as that for a dilute gas (eqn (2.82)). However, note that q in eqn (2.170) includes the influence of the solute on the solvent. Since the free energy of the solvent in eqn (2.170) does not influence the N_2 dependence, it can be omitted in most problems (or we can use a scale of free energy for solutions with the pure solvent taken as zero).

In obtaining eqn (2.170), we can use the analogy with an ideal gas: we integrate the N_2 particles over all possible configurations and for each configuration of solute particles we then integrate the solvent particles over all configurations. Since the solute particles are very far apart (dilute solution) the average value of the integration of the solvent particles will be independent of the solute positions.

2.2. Intermolecular forces

The kinetic energy of a particle is given by $\tfrac{1}{2}mv^2$ (or $p^2/2m$), this being the case for all particles in classical mechanics and quantum mechanics. What makes one physical situation different from another is the form of the potential energy. In this section I will outline the basic types of potential energy that occur in molecular systems.

2.2.1. Electrostatic energy

The basic potential energy in molecular systems is given by Coulomb's law which in practical units is

$$\varphi(r) = 139 q_1 q_2 / r, \qquad (2.172)$$

where $\varphi(r)$ is in kJ mol^{-1}, r in nm, and q in units of the electronic charge (an electron has $q = -1$, a proton $q = +1$). For $q_1 = +1$, $q_2 = -1$, column A in Table 2.1 gives $\varphi(r)$ as a function of r; for $r = 0.3$ nm the energy of interaction between unlike unit charges *in vacuo*, is large, -460 kJ mol^{-1}, of the order of a strong covalent bond.

To illustrate the use of eqn (2.172) in atomic systems we

TABLE 2.1

Electrostatic potential energy between opposite unit charges as a function of distance (nm) and relative permittivity (energy in kJ mol^{-1})

r(nm)	A Water D=78.54	B Methanol D=32.63	C Octane D=1.95	D Vacuum D=1.00
0.10	-17.56	-42.22	-706.8	-1379
0.25	- 7.02	-16.93	-282.6	- 552
0.50	- 3.51	- 8.44	-141.3	- 276
1.00	- 1.76	- 4.22	- 70.6	- 138
1.50	- 1.12	- 2.84	- 47.2	- 92
2.00	- 0.88	- 2.13	- 35.5	- 69

use the hydrogen atom as a simple example. Quantum mechanics tells us that the average distance of the electron from the nucleus (proton) is a_0 = 0.0529 nm. Then an estimate of the average potential energy in the hydrogen atom is

$$\langle E_{pot} \rangle \sim -138.8/0.0529 = -2624 \text{ kJ mol}^{-1}. \tag{2.173}$$

Eqn (2.173) happens to give the average potential energy exactly since for the 1s electron (ground state) in hydrogen

$$\psi_{1s} = \exp(-r/a_0)$$

$$\langle E_{pot} \rangle = \frac{-138.8 \int_0^\infty \psi_{1s}^2 \left(\frac{1}{r}\right) 4\pi r^2 dr}{\int_0^\infty \psi_{1s}^2 \, 4\pi r^2 dr} = \frac{-138.8}{a_0}.$$

From the virial theorem we have in general

$$\langle \text{net energy} \rangle = \tfrac{1}{2} \langle E_{pot} \rangle = -\langle E_{kin} \rangle,$$
$$\langle \text{net energy} \rangle = -1310 \text{ kJ mol}^{-1}, \tag{2.174}$$
$$\langle E_{kin} \rangle = +1310 \text{ kJ mol}^{-1}.$$

The point here is that eqn (2.172) and the virial theorem allow us to easily understand the energy of the ground state of the hydrogen atom. Eqn (2.173) also gives an extreme value of the potential energy to be found in chemical systems since a_0 = 0.0529 nm (the Bohr radius) is as close (on the average) as unlike unit charges come.

Eqn (2.172) states that two charges interact as r^{-1}. We now want to show how linear combinations of terms varying as r^{-1} give a function of r that does not vary simply as r^{-1}. Consider first the charge configuration shown in Fig. 2.1(a), where a charge q is at a variable distance r from the charges $-\delta$ and $+\delta$ which are separated by a constant distance d. The total energy is a sum over all pairs of variable Coulombic interactions,

$$\varphi(r) = 138.8\left\{\frac{q\delta}{(r+d)} - \frac{q\delta}{(r-d)}\right\}, \qquad (2.175)$$

which can be rearranged to give

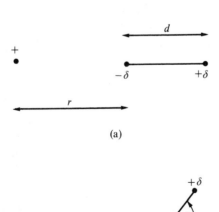

(a)

(b)

FIG. 2.1. Geometry of the interaction of a unit positive charge with a neutral charge doublet.

$$\varphi(r) = \frac{-138.8 \; q(\delta d)}{r^2\{1 + (d/r)\}}. \tag{2.176}$$

The definition of the dipole moment μ (not to be confused with the chemical potential) is

$$\mu = 48\delta d, \tag{2.177}$$

where d is in nm, and δ is in units of the electronic charge. The numerical factor 48 gives μ in debye units; unit charges separated by 0.1 nm have a dipole moment of 4.8 debyes. With the definition of μ in eqn (2.177), eqn (2.176) becomes

$$\varphi(r) = \frac{-(138.8/48)q\mu}{r^2(1 + (d/r))}. \tag{2.178}$$

If $r \gg d$

$$\varphi(r) \sim \frac{-(138.8/48)q\mu}{r^2}. \tag{2.179}$$

Thus the interaction of a charge and a dipole varies as r^{-2} as the separation between them becomes large; note that the r^{-2} dependence was derived from a linear combination of r^{-1} (Coulombic) terms (eqn (2.175)). If the dipole is not colinear with the charge, as in Fig. 2.1(b), then we easily find

$$\varphi(r) \sim \frac{-(138.8/48)q\mu \cos \vartheta}{r^2}. \tag{2.180}$$

Throughout this chapter we will include numerical factors such as the 138.8 and 48 in eqns (2.179) and (2.180) since our purpose is to estimate energies; eqns (2.179) and (2.180) give the energy in kJ mol^{-1} since they were derived from eqn (2.172).

Table 2.2 gives the dipole moments of common molecules[8] in debye units. As an illustration of the interaction of a charge with a dipole we consider the attractive interaction of a water molecule, μ = 1.86, with a positive charge (e.g. Na$^+$). Then for $r = 0.3$ nm, eqn (2.179) gives

$$\varphi(r) \sim \frac{-(138.8/48)(+1)(1.86)}{0.3^2} \simeq -60 \text{ kJ mol}^{-1}. \tag{2.181}$$

TABLE 2.2

Dipole moments of some simple molecules

Molecule	(μ debyes)
HF	1.91
HCl	1.03
HBr	0.80
HI	0.42
H_2O	1.86
methanol	1.70
ethanol	1.70
NH_3	1.47
methylamine	1.24
ethyl ether	1.15
acetic acid	1.74
methyl acetate	1.72
acetaldehyde	2.72
nitromethane	3.44

As we will discuss shortly, the charge distribution of any molecule is only approximately described by the dipole moment. None the less, eqn (2.180) does contain the major part of the potential energy of interaction between a charge (ion) and a polar (uncharged) molecule. The binding energies of water molecules to alkali metal ions in the gas phase have been measured using mass spectrometry. Table 2.3 compares the total experimental energy[9] of interaction of an ion with a water molecule with that calculated from eqn (2.179) using $r = r_{ion} + r_{H_2O}$, r_{ion} and r_{H_2O} being the radii of the respective species.

Next consider the charge configuration shown in Fig. 2.2, the collinear interaction of two dipoles. The net energy again is the sum of all variable Coulombic terms (d_1 and d_2 are fixed):

$$\varphi(\ ) = 138.8 \left\{ \frac{\delta_1 \delta_2}{(r+d_1)} - \frac{\delta_1 \delta_2}{r} + \frac{\delta_1 \delta_2}{(r+d_2)} - \frac{\delta_1 \delta_2}{(r+d_1+d_2)} \right\}. \quad (2.182)$$

Placing all terms over a common denominator and dropping terms

TABLE 2.3
Energies of formation of water–ion bonds (energies in kJ mol^{-1})

Ion	Experimental	Calculated[†]
Li^+	-142.1	-113
Na^+	-100.3	- 88
K^+	- 74.8	- 67
Rb^+	- 66.5	- 59
Cs^+	- 57.3	- 54
F^-	- 97.4	- 67
Cl^-	- 54.8	- 50
Br^-	- 52.7	- 46
I^-	- 41.8	- 38

[†]Using eqn (2.179) with μ_{H_2O} = 1.86, r_{H_2O} = 0.15 nm, and the ion radii in the order listed: 0.068, 0.098, 0.133, 0.148, 0.167, 0.133, 0.181, 0.196, 0.219 nm.

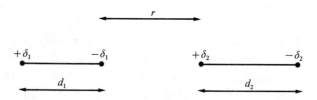

FIG. 2.2. Geometry of the collinear interaction of two neutral charge doublets.

containing d_1/r and d_2/r in the limit that both d_1 and d_2 are much greater than r,

$$\varphi(r) \simeq \frac{-2(138.8)(\mu_1/48)(\mu_2/48)}{r^3}, \qquad (2.183)$$

using the definition of μ in eqn (2.177). Thus the collinear interaction of two dipoles varies like r^{-3} at large separation. For the attractive interaction of two HF molecules ($\mu_1 = \mu_2 = 1.91$ debyes) at a distance of 0.255 nm, eqn (2.183) gives

$$\varphi(r) = \frac{-2(138.8)(1.91/48)^2}{(0.255)^3} \simeq -25 \text{ kJ mol}^{-1}. \qquad (2.184)$$

This energy is typical of the interaction of two dipoles at ~0.3 nm and is seen to be of the order of magnitude of the hydrogen bond (to be discussed shortly). HF is known to dimerize in the gas phase with a dimer energy[10] of formation of about -25 kJ mol^{-1}. Using the known distances,[11] 0.1 nm for the FH bond distance and 0.255 nm for the F ... F distance in the dimer (giving $d_1 = d_2 = 0.1$ and $r = 0.155$ for the configuration in Fig. 2.2) and the charges $\delta_F = -0.398$, $\delta_H = +0.398$ (obtained from the dipole moment and bond distance using eqn (2.190), eqn (2.182) gives φ (0.155) = -31.8 kJ mol^{-1}, which is seen to reproduce the dimerization energy (we have calculated only the attractive energy; there will also be some repulsive energy at the distance of closest approach) and is of the same magnitude as given by eqn (2.184).

The interaction of two point dipoles with arbitrary orientation is given by

$$\varphi(r) = \frac{\mu_1 \cdot \mu_2 - 3(\mu_1 \cdot \hat{r})(\mu_2 \cdot \hat{r})}{r^3}, \qquad (2.185)$$

where $\hat{r}$ is a unit vector directed along the line of centres of the two dipoles. The energies of typical orientations of dipoles are (leaving out the numerical factor): head-to-tail, $\varphi = -2\mu^2/r^3$; tail-to-tail or head-to-head, $\varphi = 2\mu^2/r^3$; perpendicular, $\varphi = 0$. For the head-to-tail arrangement, eqn (2.185) reduces to eqn (2.183).

In electrostatic theory[12] it is shown that a general charge distribution $\rho(\mathbf{r})$ can be characterized by the moments of the distribution, the integrals being over the volume of the charge distribution

$$\text{(net charge)} \quad q = \int \rho(\mathbf{r}) d\mathbf{r},$$
$$\text{(dipole moment)} \quad \boldsymbol{\mu} = \int \mathbf{r}\rho(\mathbf{r}) d\mathbf{r}. \qquad (2.186)$$

Note that q is a scalar while $\boldsymbol{\mu}$ is a vector. The next two moments are respectively the quadrupole and octupole moments. Then two arbitrary charge distributions with moments q_1, $\boldsymbol{\mu}_1$,

... and q_2, μ_2, ... have the electrostatic potential energy (leaving out the numerical factors of eqns (2.183), (2.180), and (2.172))

$$\varphi_{12} = \frac{q_1 q_2}{r} - \frac{\mu_1 \cdot \hat{r}}{r^2} - \frac{\mu_2 \cdot \hat{r}}{r^2} + \\ + \frac{\mu_1 \cdot \mu_2 - 3(\mu_1 \cdot \hat{r})(\mu_2 \cdot \hat{r})}{r^2} + \ldots \quad (2.187)$$

The various moments of the charge distribution (eqn (2.186)) are known as multipoles (monopole, dipole, quadrupole, octupole, etc.). The potential energy for the interaction of two arbitrary charge distributions is seen to involve terms representing all possible interactions between the respective multipoles. In general, the interaction between an n-pole and an m-pole has the r dependence $r^{-(n+m-1)}$. To emphasize the above points, we rewrite eqn (2.187) schematically showing only the r dependence and the nature of the multipole product:

$$\begin{aligned}\varphi_{12} = &\ (1/r) \quad + (1/r^2) \quad + (1/r^2) \quad + (1/r^3) \\ &\text{monopole-} \quad \text{monopole-} \quad \text{dipole-} \quad \text{dipole-} \\ &\text{monopole} \quad \text{dipole} \quad \text{monopole} \quad \text{dipole} \\ + &\ (1/r^3) \quad + (1/r^3) \quad + (1/r^4) \quad + (1/r^4) \\ &\text{monopole-} \quad \text{quadrupole-} \quad \text{dipole-} \quad \text{quadrupole-} \\ &\text{quadrupole} \quad \text{monopole} \quad \text{quadrupole} \quad \text{dipole} \\ + &\ (1/r^5) \quad + \ldots \\ &\text{quadrupole-} \\ &\text{quadrupole} \end{aligned} \quad (2.188)$$

If we include an infinite number of terms then eqn (2.187) gives the exact φ_{12}. In general, the first non-zero multipole product dominates φ_{12}, i.e. the series converges rapidly. Because of the inverse powers of r, higher terms become more negligible as r increases.

For the interaction of charged and/or polar molecules, an approach that is more amenable to computation than eqn (2.187) is to replace the continuous charge distribution $\rho(\mathbf{r})$, which in principle can be calculated from quantum mechanics, by a set of discrete charges (monopoles). In this approximation the potential energy of two charge distributions is given by a sum over Coulombic terms between all pairs of monopoles

$$\varphi_{12} = \sum_i \sum_j q_i q_j / r_{ij} , \qquad (2.189)$$

where the sums over i and j are respectively over the monopoles representing each charge distribution. A set of charges also can be expressed in terms of moments; thus we can choose the q_i and q_j such that eqns (2.189) and (2.187) give identically the same first few moments (in practice only the net charge and dipole moment are known for most molecules).

It is common practice to place the charges q_i on the centres of atoms in a molecule; this is simply a convenient choice and may not necessarily be the best choice to represent the charge distribution of the molecule. For example, from Table 2.2, the dipole moment of HCl is 1.03 debyes. The bond distance in this molecule is 0.127 nm. Using eqn (2.177) we have

$$\delta = \frac{\mu}{(48)d} = \frac{1.03}{(48)(0.127)} = 0.17. \qquad (2.190)$$

Thus, if the charges + 0.17 and - 0.17 are placed respectively on the H and Cl nuclei, this set of two point charges will reproduce the experimental dipole moment of HCl exactly. We should note that we customarily call such a pair of equal and opposite charges separated by a finite distance a dipole. There is a difference between such a discrete charge doublet and the point dipole of eqn (2.186): the discrete doublet has a dipole moment described by eqn (2.177) but also has higher moments (quadrupole, etc.); a point dipole has only a dipole moment and is defined as

$$\mu = \lim_{\substack{d \to 0 \\ \delta \to \infty}} (\delta d)$$

such that δd remains constant. A point dipole can also be thought of as the derivative of a delta function. The same remarks apply to all multipoles (except a monopole, which is always a point charge). Thus to a first approximation the charge distribution in CO_2 can be represented as a discrete linear quadrupole

$$\begin{array}{ccc} -\delta & +2\delta & -\delta \\ O = & C & = O \end{array}$$

Such a discrete set of charges will have zero net charge and dipole moment, a quadrupole moment, and also higher moments; a point quadrupole will have only a quadrupole moment.

Fig. 2.3 shows approximate charge distributions in formamide and methyl formate using atom-centred charges.[13] The charge distributions are based on semi-empirical quantum-mechanical calculations and experimental data on the magnitude and direction of the dipole moment (the charge distribution shown

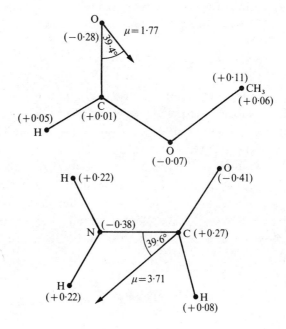

FIG. 2.3. Approximate charge distributions in formamide and methyl formate.[13] The charge distributions shown reproduce the known magnitude and direction of the dipole moments (which are shown).

reproduces both the magnitude and direction of the experimental dipole moment). There are many different approximations for obtaining charge distributions in molecules. I do not want to discuss these methods here in detail but simply want to indicate the order of magnitude of the atomic charges in a few important groups. The important point to remember is that the use of atom-centred charges is an arbitrary though useful choice; however, a set of monopoles can be introduced that reproduces an arbitrary number of the experimental moments of any charge distribution. Unfortunately, we generally know experimentally only the dipole moments and must rely on approximate quantum-mechanical calculations for more details of the charge distribution.

I wish to close this section by remarking that the electrostatic interaction of complex molecules is simple in principle, involving the summation over Coulombic terms between point charges or volume elements of charge density. Very accurate calculations await accurate detailed quantum-mechanical calculations of charge densities in large molecules. For such interactions as charge—charge, charge—dipole, and dipole—dipole the equations given in this section serve to give the potential energy with an error of the order of 10 per cent. Thus the relations given here serve to accurately reflect the magnitude of one of the most important sources of potential energy in molecular systems, simple electrostatic interactions.

2.2.2. *Polarizability of molecules*

In the preceding section we have assumed that the charge distribution of a molecule is constant and unchanged by the approach of another molecule. In fact the electric field of one charge distribution will tend to alter the charge distribution of another and vice versa. Consider a neutral atom placed in an electric field **E**. The electric field will tend to polarize the atom, moving the centre of gravity of the positive charge (nucleus), thus producing a dipole moment in the neutral atom. To first approximation this effect will be linear in the strength of the electric field (the induced moment being in the direction of the field),

$$\mu_{ind} \sim E,$$

or introducing the proportionality constant α (the polarizability) we have

$$\mu_{ind} = \alpha E. \qquad (2.191)$$

The work needed to create an induced dipole moment is

$$\varphi = -\int_0^E \alpha E dE = -\tfrac{1}{2}\alpha E^2. \qquad (2.192)$$

Eqn (2.192) gives the potential energy of a neutral atom with an induced dipole moment in the present of an electrical field strength E. The electric field can be imposed externally (through capacitor plates) or can arise from the presence of a charged or polar molecule nearby. Charged and polar molecules respectively have the following fields at a distance r (recall that the field is the negative gradient of the electrostatic potential, the potential energy of a test charge being the charge times the electrostatic potential):

field

$$\text{charge} \qquad E = q/r^2 \qquad (2.193)$$

$$\text{dipole} \qquad E = 2\mu \cos\vartheta / r^3.$$

Then the form of the field strength E given in eqn (2.193) when used in the general relation (eqn (2.192)) for the energy of a neutral, polarizable atom in a field gives

$$\begin{array}{ll}\text{(charged-induced dipole)} & \varphi(r) = -\tfrac{1}{2}\dfrac{\alpha q^2}{r^4} \\[1em] \text{(permanent dipole-induced dipole)} & \varphi(r) = -\dfrac{2\alpha\mu^2 \cos^2\vartheta}{r^6}.\end{array} \qquad (2.194)$$

Introducing the numerical factors used earlier, in practical units eqns (2.194) are

(charged-induced dipole)

$$\varphi(r) = -\tfrac{1}{2}\left(\frac{\alpha}{r^3}\right)\frac{(138.8)q^2}{r} \qquad (2.195)$$

(permanent dipole-induced dipole)

$$\varphi(r) = -2\left(\frac{\alpha}{r^3}\right)\frac{(\mu/48)^2(138.8)\cos^2\vartheta}{r^3}$$

The potential energy is given in kJ mol^{-1} if q is in units of the electronic charge, r in nanometres, and μ in debyes. The polarizability α has the dimensions of nanometres^3 (volume polarizability); hence α/r^3 is dimensionless.

Table 2.4 lists the polarizabilities of common molecules. Note that a polyatomic molecule, while in general polarizable, will not be isotropically polarizable as neutral atoms are. Thus in general the potential energy of a molecule due to induced dipole moments will depend on the direction of the field with respect to the molecule; in particular, molecules are more polarizable along covalent bonds than in the perpendicular direction. The α for polyatomic molecules in Table 2.4 thus represents an average over all orientations of the molecule. An important property of the polarizability is that it tends to increase with molecular weight, reflecting the fact that electrons further from the nucleus are more influenced by an electric field.

To illustrate eqn (2.195), consider the interaction of an argon atom with a Na$^+$ ion at a distance of 0.3 nm. Eqn (2.195)

TABLE 2.4

Polarizabilities of some molecules, atoms, and ions

Species	α(nm^3)	Species	α(nm^3)
He	0.000205	H_2O	0.00148
Ne	0.00039	NH_3	0.00224
Ar	0.00164	O_2	0.00157
Kr	0.00248	N_2	0.00174
HCl	0.00263	Na	0.0215
HBr	0.00358	Na$^+$	0.000255
HI	0.0054	Cl$^-$	0.002974

gives

$$\varphi(0.3) = -\tfrac{1}{2}\left(\frac{1.64 \times 10^{-3}}{0.3^3}\right)\left(\frac{138.8\ (1)^2}{0.3}\right) = -14\ \text{kJ mol}^{-1}. \quad (2.196)$$

The potential energy of a charge-induced dipole interaction, -14 kJ mol^{-1}, is seen to be about the same magnitude as that between permanent dipoles at the same distance (eqn (2.184)). The potential energy of a permanent dipole and an induced dipole is much weaker.

While generally overshadowed by electrostatic interactions (charge–charge, charge–dipole), field-induced dipole potential energies are by no means negligible for fields produced by unit charges. If we approximate the charge distribution of a molecule by a set of point charges as described in the previous section then a molecule produces a net field at a point outside the molecule which is the vector sum of terms like q_1/r^2 (see eqn (2.193)). This net field can then be used in eqn (2.192) to calculate the interaction of an atom with polarizability α at this point.

2.2.3. *Statistical potentials*

The interaction of a charge with a dipole or between two dipoles (eqns (2.180) and 2.185) respectively) depends on the mutual orientation of the species. In the gas phase or in solution the orientation is not necessarily fixed. Thus we are led to the notion of a statistical potential which gives the potential energy at a given distance averaged over all orientations. In a dilute gas we may use a simple Boltzmann averaging process. Using eqn (2.180), the Boltzmann averaged potential energy between a charge and a dipole is given as

$$\langle \varphi(r) \rangle = \frac{\int_0^\pi -\frac{\mu}{r^2} \cos\vartheta \exp(\mu \cos\vartheta/r^2 kT) \sin\vartheta\, d\vartheta}{\int_0^\pi \sin\vartheta\, d\vartheta}. \quad (2.197)$$

Letting

$$x = \mu/r^2 kT, \quad (2.198)$$

eqn (2.197) is given by

$$\langle \varphi(r) \rangle = -\frac{\mu}{r^2} L(x), \qquad (2.199)$$

where

$$L(x) = \coth(x) - 1/x$$
$$= \frac{x}{3} - \frac{x^3}{45} + \ldots \qquad (2.200)$$

The function $L(x)$ is known as the Langevin function and approaches unity as x becomes large. If $x < 1$ (i.e. $\mu/r^2 < kT$) then we can approximate $L(x)$ in eqn (2.200) by the leading term $x/3$, giving

(charge–dipole) $\qquad \langle \varphi(r) \rangle \simeq - \mu^2/3kT\, r^4. \qquad (2.201)$

Thus the potential energy for a charge and a dipole averaged over all orientations varies as r^{-4} for $\mu/r^2 < kT$; this is to be compared with the variation as r^{-2} (eqn (2.180)) for a fixed orientation. In solution where there are other surrounding molecules (obviously influencing the rotation of the dipole) eqn (2.197) does not accurately describe the potential energy, though it will still be true in this case that the average potential will fall off more rapidly than r^{-2} (i.e. some averaging will take place).

For two permanent dipoles we find, following a similar procedure (Boltzmann averaging over all orientations at fixed r, keeping only the leading term for $\mu/r^3 < kT$),

(dipole–dipole) $\qquad \langle \varphi(r) \rangle \simeq - 2\mu^4/3kT\, r^6. \qquad (2.202)$

2.2.4. van der Waals interactions

There is an attractive potential energy between nonpolar molecules such as the hydrocarbons, as is attested by the fact that such molecules can exist in condensed phases (liquid or solid). The origin of the potential energy is not direct electrostatic interaction, but does have its origins in Coulomb's law. In second-order perturbation theory in quantum mechanics it is found that there is a tendency for an instantaneous electronic configuration in an atom whose charge distribution is on the average symmetric to induce a dipole moment in another neutral atom, this interaction being the mathematical analogue of an instantaneous dipole-induced dipole. As with a permanent dipole-induced dipole interaction (eqn (2.194)), the leading term in this interaction is found to vary as r^{-6},

$$\varphi(r) = -C_6/r^6, \tag{2.203}$$

where C_6 is given by

$$C_6 = \frac{3}{2}\left\{\frac{(P_A)(P_B)}{P_A+P_B}\right\}\alpha_A\alpha_B, \tag{2.204}$$

where P is the ionization potential and α is the polarizability. Eqn (2.204) was first derived by London, and the interactions described by eqn (2.203) are often called London forces[14] (strictly, the force is the negative gradient of the potential); the term van der Waals interaction comes from the inclusion by van der Waals of an empirical term reflecting the existence of attractive interactions between neutral molecules in an equation of state bearing his name.

Note that eqn (2.203) gives the leading term in a second-order perturbation calculation. While r^{-6} is the dominating term, there are weaker dependences such as r^{-8}, r^{-10}, etc. Thus C_6 given by eqn (2.204) gives only an estimate of the magnitude of the van der Waals interaction. It is common practice to treat C_6 as an empirical parameter to be determined experimentally; we will give some examples of how this is done shortly.

To illustrate the order of magnitude of van der Waals interactions we calculate C_6 for the interaction of two argon atoms using eqn (2.204). The first ionization potential of argon is 1519 kJ mol^{-1} with the polarizability being given in Table 2.4 (note that α^2/r^6 is dimensionless). We have

$$C_6 = \frac{3}{2}(1519)(1.64\times10^{-3})^2 = 0.0061 \text{ kJ nm mol}^{-1}.$$

At a distance of 0.384 nm the potential energy using eqn

(2.203) is

$$\varphi(0.384) = -\frac{0.0061}{(0.384)^6} = -\frac{0.0061}{3.2\times 10^{-3}} = -1.91 \text{ kJ mol}^{-1}. \quad (2.205)$$

Eqn (2.205) gives the typical order of magnitude of the van der Waals interactions, several kJ mol^{-1}.

2.2.5. *Repulsive potentials*

So far we have been discussing attractive interactions between molecules. It is clear that the net potential between two molecules can be attractive only up to a minimum distance, the potential becoming repulsive at very short distances to prevent overlap of the molecules. The origin of the repulsive potential is quantum mechanical, though not strictly of Coulombic nature. The Pauli exclusion principle states that no two electrons in an atom can have the same set of quantum numbers. As two atoms are brought together, the only way the electrons of one can be accommodated on the other is to partially promote the electrons to higher energy levels, resulting in a large positive increase in the energy. The increase in energy is in first approximation proportional to the electron overlap as the atoms are brought together. Since the hydrogen wavefunctions are all modulated by an exponential form, the repulsive potential is commonly expressed as a simple exponential

$$\varphi(r) = A \exp(-ar). \quad (2.206)$$

Since the origin of eqn (2.206) is only semi-theoretical, the parameters A and a are obtained empirically. The determination of A and a requires the total potential for an interaction, containing both attractive and repulsive terms, a topic to which we now turn.

2.2.6. *Simple net potentials*

To obtain a net potential for a neutral-atom—neutral-atom interaction we add the repulsive and attractive potentials of eqns (2.206) and (2.203) respectively and obtain

$$\varphi(r) = A \exp(-ar) - C_6/r^6. \quad (2.207)$$

Eqn (2.207) must be considered a theoretical first approximation to the interaction between two neutral atoms if we are considering the *a priori* computation of A, a, and C_6. However, eqn (2.207) is best viewed as giving a theoretically reasonable dependence of $\varphi(r)$ on r, the parameters A, a, and C_6 being determined empirically such that $\varphi(r)$ gives exactly a number of experimentally determined properties. Note that if we determine A, a, and C_6 empirically to give exactly a set of experimental quantities, it would be quite inconsistent to add theoretical improvements to eqn (2.207) after A, a, and C_6 are so determined. Of course if we are computing a potential *a priori* from quantum mechanics then the potential is improved with added refinements.

Two simple experimental quantities that can be used to determine A, a, and C_6 in eqn (2.207) are the depth and position of the minimum in φ, $-\varepsilon$, and r_m can be exchanged for two of the parameters in eqn (2.207) using the relations

$$\varphi(r_m) = -\varepsilon,$$

$$\left(\frac{\partial \varphi}{\partial r}\right)_{r=r_m} = 0. \qquad (2.208)$$

Choosing to eliminate A and C_6 we obtain

$$A/C_6 = \frac{6}{a r_m^7} \exp(a r_m),$$

$$C_6/r_m^6 = \frac{\varepsilon}{1 - (6/a r_m)}, \qquad (2.209)$$

$$\varphi(r_m) = \frac{\varepsilon}{\left\{1 - \left(\frac{6}{a r_m}\right)\right\}} \left\{\frac{6}{a r_m} \exp\{-a(r - r_m)\} - \left(\frac{r_m}{r}\right)^6\right\}.$$

The potential of eqn (2.207), or the practical form of eqn (2.209), is often referred to as the Buckingham or '6-exp' potential.

A practical simplification of eqn (2.207) results from replacement of the exponential form of the repulsive term by an inverse power of r. In general a potential of the form

$$\varphi(r) = A/r^t - C/r^s \quad (t > s) \tag{2.210}$$

can be written in terms of ε and r_m using eqn (2.208), giving

$$A/C = \frac{s}{t} r_m^{t-s},$$

$$C/r_m^s = \frac{\varepsilon}{1-(s/t)}, \tag{2.211}$$

$$\varphi(r) = \frac{\varepsilon}{\{1-(s/t)\}} \left\{ \frac{s}{t}\left(\frac{r_m}{r}\right)^t - \left(\frac{r_m}{r}\right)^s \right\}.$$

A convenient though arbitrary choice of t for $s = 6$ is $t = 12$, giving the '6–12' or Lennard–Jones potential,

$$\varphi(r) = \varepsilon \left\{ \left(\frac{r_m}{r}\right)^{12} - 2\left(\frac{r_m}{r}\right)^6 \right\}. \tag{2.212}$$

For $s = 6$, eqn (2.211) and eqn (2.209) can be forced to have the same ε, r_m, and attractive potential (C_6 term) with the condition

$$\left(1 - \frac{6}{ar_m}\right) = \left(1 - \frac{6}{t}\right) \tag{2.213}$$

or

$$t = ar_m. \tag{2.214}$$

With the same ε, r_m, and C_6, the '6–12' potential gives a much steeper repulsive part of the potential than the '6-exp' potential.

The parameters ε and r_m in eqn (2.211) (and a in eqn (2.209)) can be determined from experimental data; in particular, as we will discuss shortly, ε and r_m can be obtained f simple substances from the heat of sublimation (obtained from calorimetry) and the distance of closest approach (obtained from X-ray studies) in the crystal. Another source of information about the potential comes from the study of gases. The equation of state of a real gas may be expressed in a power series in the density

$$p/\rho kT = 1 + B_2\rho + B_3\rho^2 + \ldots, \qquad (2.215)$$

where B_2, B_3, etc. are referred to as the second, third, etc. virial coefficients respectively. In statistical mechanics it is shown that in principle the virial coefficients can be calculated from the intermolecular potential $\varphi(r)$. The second vital coefficient is given by the relation

$$B_2(T) = 2\pi \int_0^\infty \{1-\exp(-\varphi(r)/kT)\}r^2 dr. \qquad (2.216)$$

$B_2(T)$ is obtained experimentally simply by measuring the pressure of a gas as a function of temperature and density, the slope of $p/\rho kT$ versus ρ as ρ goes to zero being $B_2(T)$. In principle, three experimental values of $B_2(T)$ are sufficient to determine three unknown parameters (ε, r_m, and a) in $\varphi(r)$. In practice, $B_2(T)$ is not sensitive to the details of the potential, being well reproduced by any potential having the correct ε (depth of the potential) and area for the attractive part of the potential (reflecting the range of the potential).

Table 2.5 gives empirical parameters for both the '6-exp'

TABLE 2.5

Approximate parameters for '6-exp' and '6-12' potentials

Atom pair	ε (kJ mol^{-1})	r_m (nm)	a (nm^{-1})
He–He	0.084	0.287	50.5
Ne–Ne	0.293	0.312	46.1
Ar–Ar	1.003	0.382	36.2
Kr–Kr	1.430	0.404	30.3
H–H	0.050	0.240	
C–C	0.501	0.340	
N–N	0.711	0.310	
O–O	1.003	0.304	

(eqn (2.209)) and '6–12' (eqn (2.212)) potentials for various substances. Since any atom, whether in a molecule or not, has both a van der Waals attractive potential with any nearby atom (again, in a molecule or not) and a repulsive potential at close enough distances, every atom pair will to first approximation have a net potential like eqn (2.207). Eqn (2.207), representing an isotropic, atom-centred potential, is a first approximation for atoms in molecules since the polarizability in general will not be isotropic (being larger, for example, along bonds as opposed to perpendicular to bonds). None the less eqn (2.207) is commonly used for atoms in molecules, both because of computational simplicity and lack of precise data or theory to warrant a more complicated procedure. For different atom pairs we can use approximate combining rules to obtain approximate parameters. Some procedures commonly used are

$$r_m(ij) = \tfrac{1}{2}\{r_m(ii) + r_m(jj)\},$$

$$\varepsilon_{ij} = \tfrac{1}{2}(\varepsilon_{ii} + \varepsilon_{jj}), \qquad (2.217)$$

$$a_{ij} = \tfrac{1}{2}(a_{ii} + a_{jj}).$$

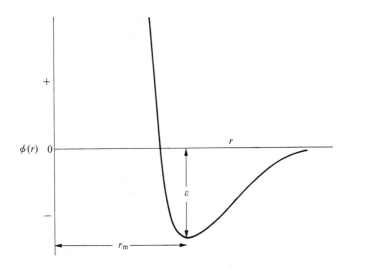

FIG. 2.4. Typical atom pair potential showing the characteristic parameters ε (depth of the well) and r_m (position of the minimum).

Eqn (2.217) represents the arithmetic mean of the appropriate parameters for the interaction between like atoms. Eqn (2.204) suggests the geometric mean for the energy, $\varepsilon_{ij} = (\varepsilon_{ii}\varepsilon_{jj})^{\frac{1}{2}}$. At any rate, eqn (2.207) is very approximate for atoms in molecules. Table 2.5 lists some parameters commonly used for atoms in biological molecules.

2.2.7. *Internal rotation potentials*

Any molecule containing four or more atoms connected in a linear fashion can exhibit different conformations due to rotation about covalent bonds (internal rotation). The classic example of internal rotation is ethane; the mode of internal rotation for this molecule is illustrated in Fig. 2.5(a). As we vary the dihedral angle ω, illustrated using only a single

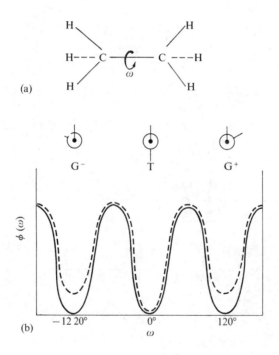

FIG. 2.5. Potential of internal rotation for ethane (solid curve); the height of the maxima is about +12.8 kJ mol^{-1}. The dotted curve shows the potential of internal rotation for butane where the minima for the two gauche states are raised by about +2.1 kJ mol^{-1}.

hydrogen atom from each methyl group in Fig. 2.5(a), the potential energy varies in a periodic fashion, being a maximum when the hydrogens are eclipsed and a minimum when they are staggered. Fig. 2.5(b) shows the nature of the internal rotation potential $\varphi(\omega)$, illustrating the fact that for ethane there are three minima and maxima, the height of the barrier being about 12.5 kJ mol^{-1}. For a symmetric potential with n equivalent maxima and minima, the simplest function that has the form shown in Fig. 2.6 is

$$\varphi(\omega) = \tfrac{1}{2}\varphi_0(1 - \cos n\omega), \qquad (2.218)$$

where φ_0 is the height of the barrier. The quantum-mechanical origin of barriers to internal rotation is still a matter of controversy.[15] We can say, however, that the barriers to internal rotation cannot be accounted for by the hydrogen–hydrogen interaction as described by eqn (2.207), i.e. the barrier is not simply caused by the hydrogen atoms on successive carbon

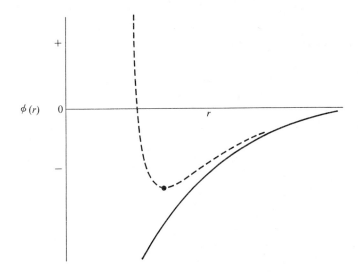

FIG. 2.6. Hydrogen-bond potential illustrating the electrostatic potential (solid curve) and the experimentally known depth and position of the minimum (dot). The dashed curve gives a net potential that includes all the known features of the potential.

atoms bumping as ω is varied. Table 2.6 gives experimentally determined barriers φ_0 for some simple molecules. We see that for ethane-like molecules the barrier height is not significantly altered by exchanging hydrogen atoms for much larger atoms such as the halogens, indicating that the barrier is primarily a basic property of the ethane molecule and not due to interactions between non-covalently bound atoms in the molecule.

2.2.8. The hydrogen bond

As with the internal rotation potential just discussed, the exact nature of the hydrogen bond[16] is a matter of controversy. Hydrogen bonding occurs between hydrogen covalently bonded to strongly electronegative atoms such as oxygen, nitrogen, or the halogens and another strongly electronegative atom. This bond is characterized by an energy of the order of -12.5 to -29 kJ mol^{-1} and an internuclear distance between the interacting hydrogen and electronegative atom that is typically about 0.1 nm shorter than the distance of closest approach as exhibited, for example, by a hydrogen in a hydrocarbon. As we have already discussed, eqn (2.183), which gives the electrostatic potential energy for the interaction of two dipoles, gives a

TABLE 2.6

Barriers to internal rotation in some simple molecules

Molecule	φ_0 (kJ mol^{-1})
CH_3CH_3	12.5
CH_3CH_2F	13.79
CH_3CHF_2	13.29
CH_3CH_2Cl	14.88
CH_3CH_2Br	14.92
CH_3OH	4.47
CH_3SH	5.27
CH_3NH_2	8.11
CH_3CHO	4.81
CH_3CO_2H	2.01

potential energy of the same order of magnitude as that found for hydrogen bonds. It is thus clear that a major part (if not all) of the energy of the hydrogen bond is simply electrostatic. If there is some component of covalent character to the hydrogen bond it must be small. Also, as we separate the groups in a hydrogen bond, any covalent bonding will disappear and the potential will be described completely by the electrostatic potential. Thus all that is known with certainty about the hydrogen bond is the distance of minimum energy r_m, the bonding energy $-\varepsilon$, and the fact that as $r > r_m$ the potential reduces to the electrostatic potential. The features just described are shown in Fig. 2.6, the dot being the position of the minimum, and the solid curve the electrostatic potential (illustrated here by eqn (2.183)); the dotted curve is a schematic representation of the net potential that incorporates all the known features of the hydrogen bond. In the next section we discuss a net potential[17] for molecules that reproduces the potential shown in Fig. 2.6.

One important feature of the basically electrostatic nature of the hydrogen bond is that the $1/r^3$ dependence for the interaction of two dipoles is not markedly sensitive to small deviations from r_m. Thus we can separate two dipoles by a few hundredths of a nanometre from r_m, or depart from collinearity by 10° to 15° at r_m and not change the energy more than a few kilojoules (out of a total energy of the order of -21 kJ).

2.2.9. *Empirical molecular potentials*

A first approximation to the net potential for molecules that includes repulsive interactions at small distances, the van der Waals attraction between all atom pairs, the electrostatic potential between atom-centred charges, and the potential due to internal rotation (if applicable) is given by

$$\varphi = \tfrac{1}{2} \sum_i \sum_j \left(A_{ij} \exp(-a_{ij} r_{ij}) - \frac{C_{ij}}{r_{ij}^6} + \frac{q_i q_j}{r_{ij}} \right) + \Sigma \varphi_{\text{int rot}}, \tag{2.219}$$

where the sum is over all atom pairs ij, the distance between which is variable either by internal rotation or simply from the fact that the atoms are different molecules. Eqn (2.219) can be used to describe the interaction between different molecules or to describe the potential energy of different conformations of a single molecule or both. The factor ½ in eqn (2.219) is important: if we sum over all values of i and all values of j each atom pair is counted twice.

For reasons I have already explained, eqn (2.219), using a sum over atom-centred pair interactions, is an approximate equation. However, the fact that the parameters are empirical and reproduce a wealth of experimental data for simple molecules (sublimation energies, distances of closest approach, and dipole moments) means that at least the major features of inter- or intramolecular interaction such as major steric interactions (atoms bumping) and major favourable or unfavourable electrostatic interactions will be described at least semi-quantitatively. Eqn (2.219) and many variations thereon[18] are much used in calculations on the conformation of biological macromolecules. This procedure is reasonable for large effects; however, numerical calculations that are based on variants of eqn (2.219) and whose significance depends on accuracy to within a few kilojoules (for a molecule) should be viewed with extreme scepticism.

As mentioned in the previous section, eqn (2.219) reproduces the known features of the hydrogen bond potential as shown in Fig. 2.6; the sum over $q_i q_j / r$ (as in eqn (2.182)) reduces to the dipole potential (eqn (2.183)) for the interaction of molecules like HCl. Of course the proper depth and position of the minimum is obtained by empirically adjusting A_{ij} and C_{ij} so as to give the known values for simple cases.

The effect of induced dipoles could be included in eqn (2.219), but greatly complicates computation since the net field at each atom must be calculated. In the spirit of the empirical nature of eqn (2.219) polarizability effects are included in that the functional form of eqn (2.219) is used to reproduce a selected set of experimental data. Thus the values of the parameters A_{ij}, etc. in eqn (2.219) reflect the presence of polarizability effects in real molecules, even though the explicit functional form appropriate for such effects is not present and must be approximated by combinations of r^{-6} and r^{-1} terms. If we include the functional form for charge-induced dipole effects in eqn (2.219) using this expanded form to obtain the parameters, we would of course obtain slightly different parameters (since the C_{ij} and q_i would not have to absorb the effects of polarizability). This raises the important point that empirical potential-energy expressions must be used in a consistent manner, i.e. a set of parameters adjusted to reproduce a set of experimental data without the explicit influence of polarizability effects are inappropriate if we introduce polarizability effects without readjusting the parameters against experimental data. These comments, of course, apply to any modification of eqn (2.219); we do not improve the energy expression by adding extra terms unless we re-determine all the empirical parameters. Empirical energy expressions are quite useless and are often very misleading unless a consistent set of parameters is used that does in fact reproduce exactly a sizeable body of data for sample systems.

2.2.10. Bond bending and stretching

In addition to the large changes in conformation that arise from internal rotation, small changes in conformation can arise by the bending and stretching of covalent bonds. These small changes can be quite important in relieving local strain or steric repulsions (bumping). To a very good approximation, the potential energy for bending and stretching can be expressed in terms of small deviations from an equilibrium bond distance and angle. Using the Hooke's law approximation of eqn (2.51b) and (2.72) we have

$$\varphi(r,\alpha) = \tfrac{1}{2}K(r-r_0)^2 + \tfrac{1}{2}K'(\alpha-\alpha_0)^2,$$

where r_0 and α_0 are respectively the equilibrium bond distance and bond angle. The force constants k and k' can be obtained experimentally from spectroscopic analysis. Table 2.7 lists stretching and bending constants for typical covalent bonds while Table 2.8 gives the energy required to stretch and bend normal carbon—carbon single bonds as a function of distance and angle. From Table 2.8 we see that while it takes a great deal of energy to stretch bonds (hence bond lengths can be treated as constant) it does not take a great deal of energy

TABLE 2.7.
Approximate stretching and bending force constants

Group	Stretching k(kJ mol^{-1} nm^{-2})	Group	Bending k'(kJ K^{-1} mol^{-1})
C—C	25 33 × 10^4	C—C—C	0.217
C=C	54 59 × 10^4		
C—N	29 33 × 10^4		
C—O	29 35 × 10^4		
C=O	71 79 × 10^4		
C—H	28 31 × 10^4		
O—H	~ 4.6 × 10^4		

TABLE 2.8.
Energy of stretching and bending carbon–carbon single bonds (energies in kJ mol^{-1})

Stretching		Bending	
Δx (nm)	$\varphi(\Delta x)$	$\Delta \alpha$ (K)	$\varphi(\Delta \alpha)$
0.001	0.13	1.0	0.11
0.005	3.14	2.5	0.67
0.010	12.54	5.0	2.71
0.015	28.01	10.0	10.87
0.020	50.16		

to bend a bond by 5° or so. If we are using eqn (2.219) to explore molecular conformations[19] in molecules that are sterically crowded or cyclic then the inclusion of bond-angle bending is essential to avoid ruling out a conformation that is excluded by large steric repulsions when using a rigid molecular skeleton but which would be allowed if some of the bond angles were allowed to relax a few degrees from their equilibrium positions.

2.2.11. *Coulombic interactions in solution*

If two point charges are placed in a homogeneous isotropic

dielectric medium with relative permittivity D, the potential energy between the charges is

$$\varphi(r) = \frac{(138.8)\, q_1 q_2}{Dr}. \qquad (2.220)$$

It is very important to note that eqn (2.220) is not a general relation of electrostatic theory but holds only for point charges in an isotropic homogeneous dielectric. Thus for ions of finite size in a solvent whose molecules also have size, shape and a nonhomogeneous electronic structure, eqn (2.220) does not strictly apply. Experimentally it is found that

$$\varphi(r) \simeq \frac{(138.8)\, q_1 q_2}{Dr} \quad (r > \text{few molecular diameters}). \quad (2.221)$$

Table 2.1 gives $\varphi(r)$ in kJ mol^{-1} as a function of r as given by eqn (2.220) for two unlike unit charges in three different solvents (octane, $D = 1.95$; methanol, $D = 32.63$; water, $D = 78.54$) and in a vacuum, where $D = 1$. From the numbers in Table 2.1 the effect of the solvent in reducing electrostatic effects is seen to be dramatic, especially in water where the interactions are reduced by almost a hundred-fold. Because electrostatic interactions are very important, the approximate nature of eqn (2.221) introduces a large uncertainty in estimating intermolecular energies in solution. The simplest empirical solution is to adopt the following form of Coulomb's law[20]

$$\varphi(r) = \frac{(138.8)\, q_1 q_2}{D(r)\, r}, \qquad (2.222)$$

where $D(r)$ has the following properties

$$D(r) = \begin{cases} 1 & \text{as } r \to 0 \\ D & \text{as } r \to \infty, \end{cases} \qquad (2.223)$$

where the transition from $D(r) = 1$ to $D(r) = D$ takes place fairly sharply when r is of the order of one or two molecular diameters.

A more satisfying approach to the interaction of two charges in solution would be to write

$$\varphi(r) = \frac{(138.8)\, q_1 q_2}{r} + \varphi(\text{charge--solvent}) + \varphi(\text{solvent--solvent})$$
(2.224)

where the charge—charge interaction is represented by a simple Coulombic interaction ($D = 1$). Eqn (2.224) is correct in form: the charges do have an attractive Coulombic interaction, the solvent—charge and solvent—solvent interaction producing a net repulsive interaction (the solvent molecules tend to concentrate in the high-field region between the charges, thus forcing the charges apart). Eqn (2.224) of course requires statistical-mechanical analysis, since we must average over all solvent configurations. Thus the simplicity of eqn (2.221), together with the fact that the single scale factor D seems to accurately represent the complicated interactions explicitly represented in eqn (2.224) as r becomes greater than a molecular diameter or so, makes eqn (2.221) very practical. We must keep in mind, however, that we are attempting to describe very complicated solvent interactions by the use of a single number characteristic of the bulk solvent. As with all empirical energy expressions, eqn (2.222) is reliable only if the form of $D(r)$ has been adjusted so as to reproduce experimental data in test cases.

2.2.12. *Hydrogen bonds in polar solvents*

The energy of the hydrogen bond as measured, for example, by the sublimation energy of crystals or dissociation of dimers in the gas phase is in the range of -21 kJ mol^{-1}. If, however, we are breaking or forming a hydrogen bond between two groups in a polar solvent, where the solvent molecules can also participate in hydrogen bonding, then the energy can be much smaller. Representing solvent molecules as SH, the formation of a hydrogen bond between a donor DH and an acceptor A group can be schematically written as an exchange of hydrogen bonds:

$$\text{DH}..\text{SH} + \text{A} \ldots \text{HS} \rightleftharpoons \text{DH} \ldots \text{A} + \text{HS} \ldots \text{HS}. \quad (2.225)$$

Depending on the relative strengths of the different bonds and the average degree of solvent bonding, the net difference in energy can be positive, negative, or very close to zero. In fact hydrogen bonds in many important biological macromolecules in water have energies of formation that are very close to zero as we will discuss in detail in later chapters.

2.2.13. Hydrophobic bonds

When argon dissolves in water the following thermodynamic parameters[21] are found at 298 K:

$$\Delta H_{\text{solution}} \simeq -16\ 720\ \text{J mol}^{-1},$$
$$\Delta S_{\text{solution}} \simeq -84\ \text{J mol}^{-1}\ \text{K}^{-1}. \quad (2.226)$$

The fact that the enthalpy of solution of argon is negative while the entropy of solution is also negative can only be reasonably explained by assuming that an argon atom increases order in the solvent (water). It is known experimentally (X-ray scattering from liquids can be used to measure the number of molecules in the first coordination shell) that liquid water retains a large amount of tetrahedral coordination (characteristic of water molecules in ice). Tetrahedral coordination leads to a very open structure into which other groups, such as an argon atom, can at least partially fit without requiring the breaking of hydrogen bonds in liquid water to accommodate the solute. Since the nonpolar solute will have an attractive van der Waals interaction with the hydrogen-bonded water molecules, the energy of hydrogen-bonded water molecules will be made more favourable. Thus, in this picture, argon stabilizes hydrogen-bonded complexes of water molecules leading to a lowering of the energy (negative enthalpy) with a concomitant lowering the entropy (the water molecules are more highly ordered). All nonpolar molecules, such as the hydrocarbons, exhibit enthalpies and entropies of solution in water similar to those given in eqn (2.226).

There have been many attempts[22] to construct a statistical-mechanical model for the interaction of water with nonpolar solutes based on the above qualitative ideas. The reader is referred to these treatments for more details.

Since all biological molecules contain hydrocarbon or hydrocarbon-like portions, the above effect on the structure of water will be present to some degree. These interactions are known as hydrophobic (water-fearing), since nonpolar molecules tend to have limited solubility in water even though the energy of solution is favourable, the negative or unfavourable entropy of solution outweighing the favourable energy. The effect of nonpolar groups on water structure has a very interesting consequence for the association of molecules in water. In qualitative terms, if we accept the view that nonpolar groups are statistically surrounded by ice-like water (i.e. containing more hydrogen bonding or tetrahedral coordination than normal water) then when the nonpolar groups are brought together there is less nonpolar surface exposed to water. Thus the amount of ordered water is reduced when two nonpolar groups are brought together, this resulting in an increase in enthalpy and entropy. The result that the association of nonpolar groups in water can be accompanied by an increase in entropy means that the associated state is favoured as the temperature is increased, hydrophobic bonds becoming weaker as the temperature is lowered. Biological molecules in general contain nonpolar portions, polar portions, and charged groups. The net thermodynamics of association will thus in general be a complex combination of hydrophobic effects, hydrogen bonds (competing with hydrogen bonds to water), and charge—charge and charge—dipole interactions.

2.3. Simple applications

Having discussed partition functions and molecular forces we now want to use some of the intermolecular potentials we have discussed in the framework of statistical mechanics to interpret the observable properties of bulk matter.

2.3.1. *Crystal potentials and sublimation energies*

One of the simplest measures of intermolecular potential energies is the energy required to transform a crystal into a very dilute gas at 0 K (sublimation energy). Table 2.9 lists experimental energies of sublimation extrapolated to 0 K for various substances.[23] As a crude first approximation, the energy of interaction between a pair of molecules at the distance of closest approach can be estimated from

TABLE 2.9

Sublimation energies (kJ mol^{-1}) for various substances at 0 K

H_2	0.79	propane	27.50
N_2	6.94	benzene	49.74
O_2	8.82	HF	11.12
F_2	8.36	HCl	20.06
Cl_2	30.18	HBr	22.91
Br_2	46.40	HI	26.17
I_2	74.00	NH_3	29.18
Ne	1.92	H_2O	47.23
Ar	7.73	H_2S	24.70
Kr	11.20	diamond	706.42
Xe	15.01	LiCl	828.06
CCl_4	41.80	NaCl	764.10
CF_4	17.39	KCl	687.19
CH_4	9.20	RbCl	670.89
ethane	20.11	Hg	64.37

$$-\varepsilon = \Delta U_{subl}/(\tfrac{1}{2} n_c), \qquad (2.227)$$

where n_c is the coordination number of number of nearest neighbours in the crystal. The coordination numbers for some of the substances listed in Table 2.9 are

substance	n_c
Ar	12
CH_4	12
H_2O	4
NaCl	6

The factor $\tfrac{1}{2}$ in eqn (2.227) prevents the counting of each pair interaction twice (note that the sublimation energy measures the energy per mole to take a macroscopic sample of solid to the gas phase and does not measure the energy to take a single molecule out of a crystal). Eqn (2.227) assumes that the potential energy between molecules extends only to nearest neighbours, while in fact every molecule in a crystal interacts, however weakly, with every other molecule. Since crystals have regular geometry we can compute the interaction of a typical molecule in a crystal with all others and thus obtain a better estimate of the intermolecular potential energy than given by eqn (2.227).

If $\varphi(r)$ is the pair potential between atoms, then the potential energy of a crystal of N atoms can be written (assuming the total energy can be expressed as a sum of atom–atom interactions) as

$$\Phi = N\left\{\tfrac{1}{2} \sum_{i=2}^{N} \varphi_{1i}(R_{1i})\right\}, \qquad (2.228)$$

where the sum is between a reference atom chosen as number one and all the other atoms in the crystal. Letting R_m be the distance of closest approach of two atoms in the crystal (as obtained from X-ray studies) then, because of the regular structure of the crystal, all the atom-pair distances R_{1i} can be expressed in terms of R_m:

$$R_{1i} = \alpha_i R_m, \qquad (2.229)$$

where α_i is the appropriate geometrical factor. Assuming that $\varphi(r)$ can be expressed in terms of powers of R, we have (illustrating only a single term in φ)

$$\varphi(R) = C/R^n, \qquad (2.230)$$

$$\tfrac{1}{2} \sum_i \varphi(R_i) = \tfrac{1}{2} \sum_i C/R_i^{\,n} = \tfrac{1}{2} \frac{C}{R_m^{\,n}} \sum_i \left(\frac{1}{\alpha_i}\right)^n.$$

The quantity $\Sigma(1/\alpha_i)^n$ represents a sum over an inverse power of the geometrical factors α_i which are characteristic of the particular crystal structure. Hence the quantity

$$M_n = \tfrac{1}{2} \sum_i \left(\frac{1}{\alpha_i}\right)^n \qquad (2.231)$$

is a constant characteristic of a given crystal (and power law, n); M_n is known as the Madelung constant.[24] Using the interatom potential of eqn (2.210), the crystal potential energy is given by

$$\Phi(R) = M_t \frac{A}{R^t} - M_s \frac{C}{R^s}, \qquad (2.232)$$

where R is the nearest-neighbour distance in the crystal. Using the relations

$$\Phi(R_m) = -\Delta U_{\text{subl}},$$

$$\left(\frac{\partial \Phi}{\partial R}\right)_{R=R_m} = 0, \qquad (2.233)$$

we have, in analogy with eqn (2.211),

$$\frac{AM_t}{CM_s} = \frac{s}{t} R_m^{t-s},$$

$$\frac{CM_s}{R_m^s} = \frac{\Delta U_{\text{subl}}}{1 - s/t}, \qquad (2.234)$$

$$\Phi(r) = \frac{\Delta U_{\text{subl}}}{(1-s/t)}\left\{\frac{s}{t}\left(\frac{R_m}{R}\right)^t - \left(\frac{R_m}{R}\right)^s\right\}.$$

Using eqns (2.234) and (2.211) we have the relations between ε and r_m (the parameters for the atom pair potential) and ΔU_{subl} and R_m (the parameters for the total potential energy of a crystal):

$$\varepsilon = \frac{\Delta U_{\text{subl}}}{M_s}\left(\frac{M_t}{M_s}\right)^{s/(t-s)},$$

$$r_m = R_m \left(\frac{M_s}{M_t}\right)^{1/(t-S)}. \qquad (2.235)$$

Table 2.10 lists the constants M_n for various values of n for crystals with hexagonal close packing[25] (the most efficient

TABLE 2.10.

Madelung constants for hexagonal close packing

n	M_n
6	7.227
8	6.401
10	6.156
12	6.066
14	6.030

way to pack hard spheres). Since in hexagonal close packing there are 12 nearest neighbours, we have $M_n' = 6$ if the sum in eqn (2.231) is extended only over nearest neighbours. Thus the amount M_n deviates from 6 is a measure of the importance of interactions beyond nearest neighbours. For the '6–12' potential ($t = 12$, $s = 6$), eqn (2.235) using the appropriate constants from Table 2.10 becomes

$$\varepsilon = \Delta U_{sub1}/(8.7),$$
$$r_m = R_m/(0.97).$$
(2.236)

We see that eqn (2.227) with $\tfrac{1}{2}n_c = 12/2 = 6$ overestimates ε noticeably. On the other hand, the minimum r_m in the pair potential is very close to the distance of closest approach R_m taken from the crystal for atoms interacting with a potential of the form of eqn (2.210) (with $t > s$), r_m being slightly larger than R_m.

For ionic crystals the procedure for calculating the crystal energy is similar to that outlined above with two differences. First, the unit in the crystal is not a single atom but the simplest formula for the salt, e.g. NaCl. Second, the sum over all interactions between a reference ion and all others involves plus and minus terms (from interactions with like and unlike ions). Taking the pair potential for the interaction of singly charged positive and negative ions as (Coulombic attraction, r^{-t} repulsion),

$$\varphi(r) = A/r^t - 138.8/r \quad (\text{kJ mol}^{-1} \text{ of ion pairs}), \quad (2.237)$$

then the energy per mole of ion pairs in a crystal having the NaCl structure (ions located on the simple cubic lattice each ion having six nearest-neighbouring ions of the opposite charge) is given by

$$\Phi(R) = M_t \, A/R^t - 1.75 \frac{(138.8)}{R}, \quad (2.238)$$

where R is the nearest-neighbour ion distance in the crystal, 1.75 is the Madelung constant for the Coulombic (r^{-1}) interaction on the NaCl-type lattice, and the factor M_t is the Madelung constant for the r^{-t} repulsive potential for the NaCl lattice (which is different from the M_n given for hexagonal close packing in Table 2.10). Using the fact that the derivative of $\Phi(R)$ is zero at $R = R_m$, we have

$$\Phi(R_m) = -1.75 \frac{(138.8)}{R_m} (1-\tfrac{1}{t}) = -\Delta U_{\text{subl}}. \quad (2.239)$$

From Table 2.9, ΔU_{subl} for NaCl is -764.1 kJ mol^{-1}, while from X-ray studies $R_m = 0.28$ nm. Thus

$$-17.5 \left(\frac{138.8}{0.28}\right)(1-\tfrac{1}{t}) = -867.5 \, (1-\tfrac{1}{t}) = -764.1. \quad (2.240)$$

Thus $t \sim 10$ gives a good fit to the experimental energy. Note that $-1.75 \times 138.8/R_m = 867.5$ kJ mol^{-1}, the attractive Coulombic energy, is about 10 per cent larger than the experimental energy, i.e. Coulombic attractions alone give a good estimate of the crystal energy.

Using eqn (2.234) and eqn (2.235) we can calculate r_m and ε for the pair potential for the NaCl interaction. We have (with $M_{10} = 6$)

$$\varphi(r) = \frac{\varepsilon}{\{1-(1/t)\}} \left\{ t^{-1}\left(\frac{r_m}{r}\right)^t - \left(\frac{r_m}{r}\right) \right\},$$

$$\varepsilon = \frac{\Delta U_{\text{subl}}}{1.75}\left(\frac{r_m}{R_m}\right) = \frac{\Delta U_{\text{subl}}}{2.01}, \quad (2.241)$$

(cont.)

$$r_m = R_m\left(\frac{1.75}{M_t}\right)^{1/(t-1)} = R_m(0.87).$$

An interesting result of eqn (2.241) is that r_m for the NaCl ion-molecule is smaller than R_m, the distance of closest approach in the crystal. This is in contrast to r_m for the interaction between neutral atoms (e.g. Ar) or roughly spherical molecules (e.g. CH_4) that crystallize in hexagonal close packing where $R_m < r_m$ (eqn (2.236)).

For crystals of polyatomic molecules eqns (2.233) do not suffice to describe static equilibrium in the crystal. The requirements of zero torque and stress on the unit cell expand eqn (2.233) to a set of 13 equations (some of which may be redundant, depending on the symmetry of the molecule). These equations can be used to determine a large number of empirical parameters in intermolecular pair potentials.[26]

2.3.2. Average distances between molecules

To obtain a first estimate of the effect of intermolecular interactions we must be able to estimate the distances between molecules. In this section I shall indicate how the average distance between molecules in liquids, solutions, or gases can be estimated from the density or concentration. For densities and concentrations in the units

$$\begin{aligned}\rho &= \text{density in kg m}^{-3}, \\ c &= \text{concentration in mol l}^{-1},\end{aligned} \quad (2.242)$$

the conversion factors

moles = g/molecular weight,
m^3 = litres × 1000,
$nm^3 = m^3 \times 10^{27}$,
number of molecules = (moles) (6.022 × 10^{23})

give

$$\text{nm}^3/\text{molecule} = \text{molecular weight}\,(0.6022\,\rho) = 1/(0.6022\,c). \quad (2.243)$$

To obtain an estimate of the average distance between molecules we must specify the arrangement of the molecules. For the arrangements

$$\begin{array}{ll} \text{cubic} & \text{nm}^3/\text{molecule} = R^3 \\ \text{hexagonal} & \text{nm}^3/\text{molecule} = R^3/\sqrt{2} \end{array} \quad (2.244)$$

where R is the distance between nearest neighbours, we have

$$\begin{array}{ll} \text{cubic} & R = (\text{nm}^3/\text{molecule})^{\frac{1}{3}}, \\ \text{hexagonal} & R = (\text{nm}^3/\text{molecule})^{\frac{1}{3}} 2^{\frac{1}{6}}. \end{array} \quad (2.245)$$

Table 2.11 gives the average distance between particles expressed in nanometres as a function of concentration expressed in moles per litre. Note that, for example, a NaCl solution of 1 M is 2 M in the number of particles.

TABLE 2.11.

Average distance between molecules as a function of concentration

c (mol l^{-1})	Cubic arrangement R(nm)	Hexagonal arrangement R(nm)
10^{-4}	25.30	28.40
10^{-3}	11.85	13.30
10^{-2}	5.45	6.12
10^{-1}	2.53	2.84
1	1.19	1.34
10	0.55	0.62

One further example is a gas at 273 K and 101 kPa pressure. Under these conditions 1 mol of an ideal gas occupies 22.4 l or the concentration is 0.0446 mol l^{-1}. This gives an average distance between molecules of 3.3 nm (assuming a cubic arrangement) or 3.7 nm (assuming hexagonal arrangement).

Of course these estimates simply give an average distance between particles, there being a broad distribution of distances in gases and dilute solutions.

2.3.3. Thermodynamics of solutions of salts

The process of dissolving a simple salt (e.g. NaCl) in water is conveniently visualized as taking place via the following steps

$$\text{salt} \xrightarrow{(1)} \text{ion gas} \xrightarrow{(2)} \text{solution in water} . \quad (2.246)$$
$$\underbrace{\hspace{4cm}}_{(3)}$$

The energy for step (1) is simply the sublimation energy of the crystal (strictly the sublimation energy at room temperature, which is not greatly different from the sublimation energy at 0 K) which from Table 2.9 is a very large positive energy (+ 764.1 kJ mol^{-1} for NaCl). The energy of step (2) can be estimated from the energy of interaction between a dipole (water molecule) and an ion. Using the energies given in Table 2.3 we can estimate the energy of dissolving ions in water by multiplying the energy of a dipole—ion pair interaction by the number of water molecules that surround an ion. Then

$$\begin{pmatrix}\text{energy of solution}\\ \text{per mole of ion}\\ \text{pairs}\end{pmatrix} = \begin{pmatrix}\text{number of waters in}\\ \text{first hydration}\\ \text{shell of cation}\end{pmatrix}\begin{pmatrix}\text{water-cation}\\ \text{pair energy}\end{pmatrix} +$$
$$+ \begin{pmatrix}\text{number of waters in first}\\ \text{hydration shell of anion}\end{pmatrix}\begin{pmatrix}\text{water-anion}\\ \text{pair energy}\end{pmatrix}. \quad (2.247)$$

A rough estimate of the number of waters in the first hydration shells for NaCl with the water—ion pair energies listed in Table 2.3 gives

$$\begin{pmatrix}\text{energy of solution}\\ \text{per mole of}\\ \text{ion pairs}\end{pmatrix} \sim (5)(-100) + (7)(-54.8) \sim -885 \text{ kJ mol}^{-1}, \quad (2.248)$$

which is seen to match the energy of sublimation (step 1) in eqn (2.246). Thus in the dissolving process, the Coulombic interactions between charges in the salt are exchanged for ion-dipole interactions in solution, these net energies being of comparable magnitudes. Also note from Table 2.12 that the enthalpy dominates the standard free energy of solution from the gas phase ($\Delta\mu^0 = \Delta g^0 = \Delta h^0 = T\Delta s^0$) an entropy change of -125 J k^{-1} mol^{-1} contributing $-T\Delta s^0 = (300)(125) \sim 40$ kJ mol^{-1}

AND INTERMOLECULAR FORCES 143

TABLE 2.12.

Thermodynamic parameters for the transfer of ions from the gas phase to water at 298 K

Salt	$\Delta\mu^0 = \Delta g^0$ (kJ mol^{-1})	Δh^0 (kJ mol^{-1})	Δs^0 (J K^{-1} mol^{-1})
NaF	-853	-911	-192
NaCl	-727	-769	-134
NaBr	-702	-736	-121
NaI	-669	-698	-100
LiCl	-831	-878	-154
KCl	-660	-690	- 96
RbCl	-635	-660	- 88
CsCl	-606	-631	- 79

to $\Delta\mu^0$ compared with $\Delta h^0 \sim 840$ kJ mol^{-1}.

While eqn (2.247) is only a very crude first estimate of the interaction of ions in water, it does accurately predict the fact that the net water–ion energy closely matches the energy in the crystal. Table 2.12 lists experimental thermodynamic parameters[27] of solution for various salts. We see that the enthalpies for reaction (3) in eqn (2.246) are small relative to the energy of sublimation of the crystals (see Table 2.9).

Eqn (2.247) estimates only the interaction of an ion with the first shell of water molecules. A very crude estimate of the energy of interaction of an ion with all the water molecules past the first shell is to use the potential for the interaction of a dipole and an ion where the interaction is averaged over all orientations of the dipole. Using the statistical potential of eqn (2.201) we have

$$U_2 = \begin{Bmatrix} \text{ion interaction} \\ \text{with water molecules} \\ \text{past first hydration} \\ \text{shell} \end{Bmatrix} = -\rho \int_{R_1}^{\infty} \frac{\mu^2}{3kT\,Dr^4}\, 4\pi r^2 dr, \quad (2.249)$$

where R_1 is the outer radius of the first hydration shell and ρ is the density of liquid water in molecules per cubic nanometre. Eqn (2.249) gives the result

$$U_2 = -\text{constant}/R_1. \tag{2.250}$$

To first approximation the energy of hydration from the first shell is independent of the radius of the ion (the volume of the first shell is approximately $4\pi R_1^2 \Delta$, where Δ is the diameter of a water molecule; since the ion-dipole interaction varies with the inverse square of the distance, the dependence of eqn (2.247) on the radius of the hydration shell cancels). This leaves the R_1 dependence of U_2 as the main effect of the size of the ion on the hydration energy. While eqn (2.249) (hence eqn (2.250)) is very approximate, the net hydration energy is found to vary inversely with ion size as, for example, in a series of salts where either the cation or anion is not varied as can be seen from the data listed in Table 2.12.

The major point we wish to make in this section is that the energy of hydration of a singly charged ion in water is very large, of the order of -400 kJ mol^{-1}. This applies not only to ions like Na$^+$ and Cl$^-$ but to charged groups like RCOO$^-$ and RNH$_3^+$. Since the total hydration energy is very large, changes in hydration due to alterations in neighbouring groups or association of the parent molecule can be accompanied by large energy changes if the state of hydration is significantly changed. For example, in the very important reaction

$$\text{ATP} \rightleftharpoons \text{ADP} + \text{P}_i, \tag{2.251}$$

the standard free energy change is -33 kJ mol^{-1}. Clearly the fact that ATP is unstable thermodynamically with respect to ADP + P$_i$ is due in part to the fact that in the triphosphate group in ATP, negative charge is concentrated in three neighbouring groups, this concentration being relieved when one phosphate splits off. However, we could easily account for all of the -33 kJ mol^{-1} of free energy by differences in the hydration energy of the triphosphate versus the diphosphate + P$_i$.

2.3.4. *Chemical potential for dilute salt solutions*

Eqn (2.87) summarizes the conditions that give rise to the law of mass action. From eqn (2.170) we have argued that dilute solutions have the form of the partition function $(V_q)^{N/N!}$ that gives rise to a chemical potential of the form (which in turn gives rise to the law of mass action)

$$\mu = \Delta\mu^0 + RT \ln c,$$

where c is the concentration. For salt solutions the quantities $\Delta\mu^0$ (the free energies of solution from the gas phase per mole of ions at infinite dilution) are given in Table 2.12 for several salts; the enthalpy contribution to $\Delta\mu^0$ having been discussed in the last section. The form of the partition function given in eqn (2.170) arises only if one can ignore any interaction between solute particles. Since the interaction between ions is Coulombic, varying as r^{-1}, which is the longest-range intermolecular potential, it is not surprising that the assumption of independent ions does not hold well even in dilute solution. It is customary to write deviations from the ideal form of the chemical potential given above in the form

$$\mu = \mu_{ideal} + \mu_{non\text{-}ideal},$$
$$\mu_{ideal} = \Delta\mu + RT\ln c, \qquad (2.252)$$
$$\mu_{non\text{-}ideal} = RT\ln \gamma(c),$$

where $\gamma(c)$ measures the deviations from ideality. In this section we want to examine the concentration dependence and magnitude of $\gamma(c)$ that arises from the long-range nature of the potential between ions. This will involve the classic Debye–Hückel theory. But first I want to give an approximate treatment that is simpler to understand than the Debye–Hückel treatment.

To estimate the electrostatic potential energy of a dilute salt solution (we will use the example of NaCl throughout this section) we can visualize the ions in solution as being arranged on a regular lattice having the same structure as the NaCl crystal,[28] the differences being that the ions are much further apart in dilute solution than in the crystal and that the space between them is filled with a solvent of high relative permitivity. Then using eqn (2.221) and the attractive Coulombic part of eqn (2.238) we have (for a total of N NaCl ion paris rather than per mole)

$$\Phi(N) = \frac{-NM_{NaCl}(138.8)}{D\langle R(N)\rangle}, \qquad (2.253)$$

where the nearest-neighbour distance $\langle R(N)\rangle$ between ions is a function of the concentration. Using eqn (2.245) for cubic packing and recognizing that a solution 1 M in NaCl is 2 M in the number of ions, we have (with units such that N/V is in moles of NaCl per litre),

$$\Phi(N) = -\left\{\frac{(138.8)M_{NaCl}}{D}\left(\frac{0.6022}{500V}\right)\right\}N^{4/3} \qquad (2.254)$$

Assuming that the Coulombic potential energy is the sole source of non-ideality, we can write

$$G = G_{ideal} + \Phi(N). \qquad (2.255)$$

Using the general definition of the chemical potential (eqn (2.25)) we have (c in moles NaCl per litre)

$$RT\ln \gamma = \frac{\partial \Phi}{\partial N} = -\frac{4/3\, M_{NaCl}(138.8\times 10^3)}{3D}\left(\frac{0.6022}{500}\right)^{\frac{1}{3}}c^{\frac{1}{3}}, \qquad (2.256)$$

where the units of energy are J mol^{-1} (using the gas constant R as 8.4 J K^{-1} mol^{-1}, the conversion from kilojoules to joules being accomplished by the factor 10^3).

Table 2.13 lists experimental[29] values of γ_+ for NaCl in water. To test eqn (2.256), $\ln \gamma_+$ is plotted versus $c^{\frac{1}{3}}$ in Fig. 2.7 ($\gamma_+ = \gamma^{\frac{1}{2}}$ is mean value of γ per ion). We see that an excellent fit is obtained in the concentration range 10^{-3} mol l^{-1} to 10^{-1} mol l^{-1}. If we treat M_{NaCl} as an adjustable parameter that is adjusted to give the slope shown in Fig. 2.7 we find M_{NaCl} = 1.14, which is to be compared with the value 1.75 for the NaCl crystal.

Clearly the Na$^+$ and Cl$^-$ ions in dilute solution are not arranged on a regular lattice. On average, however, negative ions will have positive-ion neighbours, and of course vice versa, the lattice approximation giving a very good estimate of

TABLE 2.13.

Mean ion activity coefficient for NaCl in water at 298 K as a function of concentration (eqns (2.256) and (2.260) give γ for a $Na^+ Cl^-$ pair; $\gamma_{\pm}^2 = \gamma$

$\gamma_{\pm}$	c (mol l^{-1})
0.966	0.001
0.953	0.002
0.929	0.005
0.904	0.010
0.875	0.020
0.823	0.050
0.780	0.100
0.730	0.200
0.68	0.500
0.66	1.000
0.68	2.000
0.78	4.000

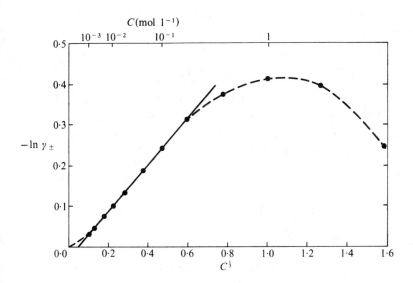

FIG. 2.7. Plot of $\ln \gamma_{\pm}$ for NaCl versus $c^{\frac{1}{3}}$. The straight-line portion illustrates the range of c over which the quasi-lattice model seems to work well.

the concentration dependence of the average distance between ions. The quasi-lattice view of dilute ionic solutions gives a very simple and surprisingly accurate account of the inter-ion potential energy. The potential energy of ions in dilute solution is given by eqn (2.254). Using M_{NaCl} obtained from fitting eqn (2.256) to experimental data gives for a 0.1 M solution of NaCl (per mole of NaCl)

$$\Phi \simeq -2.1 \text{ kJ mol}^{-1}, \qquad (2.257)$$

which is seen to be very small when compared to the hydration energy of an ion in solution discussed in the previous section.

While the quasi-lattice treatment correctly predicts the concentration dependence of $RT\ln \gamma$ (eqn (2.256)) over a wide range of concentration, the $c^{\frac{1}{3}}$ dependence does not hold for very dilute solutions. The correct concentration dependence as the concentration goes to zero is given by the Debye–Hückel theory of dilute electrolytes.

The derivation of the Debye–Hückel theory is given in any standard physical-chemistry textbook. Here I shall quote only important results. The pair potential between two ions in solution is given by

$$\varphi(r) = \frac{(138.8)q_1 q_2}{Dr} \exp(-\varkappa r), \qquad (2.258)$$

which is seen to be the same as eqn (2.221) except that the simple Coulombic potential $-(138.8)/Dr$ for point charges in a solvent with relative permittivity D is modulated by the exponential $\exp(-\varkappa r)$. The quantity $\varkappa$ for a 1–1 electrolyte at 298 K in water is

$$\varkappa = c^{\frac{1}{2}}/30.4. \qquad (2.259)$$

Since $\varkappa r$ must be dimensionless, $\varkappa^{-1}$ has the dimensions of length; eqn (2.259) gives $\varkappa$ in nanometres with c in moles NaCl per litre.

In the Debye–Hückel theory the source of nonideality, $RT\ln \gamma$ in eqn (2.252), is given by (energy in J mol^{-1}, c in moles NaCl per litre),

$$RT\ln \gamma = -\frac{(138.8) \times 10^3}{D\varkappa^{-1}} = -\frac{(138.8) \times 10^3}{30.4\ D} c^{\frac{1}{2}}. \quad (2.260)$$

Again using the data of Table 2.13, Fig. 2.8 plots the experimental values of $\ln \gamma_\pm$ versus $c^{\frac{1}{2}}$. We see that as the concentration goes to zero, eqn (2.260) predicts the density dependence of $\ln \gamma$ and gives the correct numerical value of the slope of $\ln \gamma$ versus $c^{\frac{1}{2}}$.

Comparing the Debye–Hückel theory with the quasi-lattice approach just discussed, we have (eqn (2.260) and (2.256))

$$\begin{aligned}\text{Debye–Hückel} \quad & \mu = \Delta\mu^0 + RT\ln c - (\text{constant})\ c^{\frac{1}{2}}, \\ \text{Quasi-lattice} \quad & \mu = \Delta\mu^0 + RT\ln c - (\text{constant})\ c^{\frac{1}{3}}. \end{aligned} \quad (2.261)$$

Actually the quasi-lattice approach predicts the concentration dependence of $\ln \gamma$ over a much wider range of concentration than the Debye–Hückel theory, the quasi-lattice notion breaking

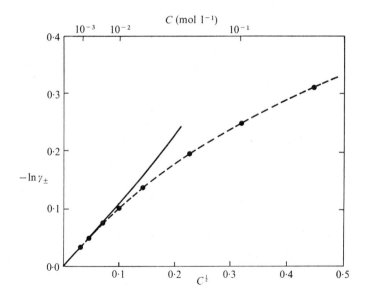

FIG. 2.8. Plot of $\ln \gamma_\pm$ for NaCl versus $c^{\frac{1}{2}}$. The straight line is the prediction of the Debye–Hückel theory.

down only for very dilute solutions. The $c^{\frac{1}{2}}$ dependence in the Debye–Hückel theory overestimates the Coulombic potential energy of the salt solution for concentrations greater than 0.01 mol l^{-1}.

In the Debye–Hückel theory the distribution of opposite charge about a central ion is shown to vary as

$$\text{charge density} \sim 4\pi r^2 \left(\frac{1}{r} \exp(-\varkappa r)\right) \sim r \exp(-\varkappa r), \qquad (2.262)$$

which has a maximum at $r = \varkappa^{-1}$. The physical picture eqn (2.262) given is a cloud of ions of opposite sign surrounding a given ion with a most probable distance of $\varkappa^{-1}$. Note that it is not inconsistent for a given ion to both have a surrounding cloud (or 'ion atmosphere') of ions of opposite charge and be in the cloud for another ion; in the NaCl crystal each Na$^+$ ion has six nearest-neighbour Cl$^-$ ions and vice versa. In the quasi-lattice view each ion is also surrounded by ions of opposite charge, the difference between the Debye–Hückel and quasi-lattice pictures being in the concentration dependence of the characteristic distance between ions. From eqns (2.253) and (2.260) we have

	Characteristic distance between charges	Concentration dependence	
Debye–Hückel	$\varkappa^{-1}$	$c^{-\frac{1}{2}}$	(2.263)
quasi-lattice	$\langle R \rangle$	$c^{-\frac{1}{3}}$	

From eqn (2.259) we have $\varkappa^{-1} = 3.04$ nm for $c = 10^{-2}$ mol l^{-1}, $\varkappa^{-1} = 30.4$ nm for $c = 10^{-4}$ mol l^{-1}; these values can be compared with $\langle R \rangle$ computed from eqn (2.245), as shown in Table 2.11.

Of particular interest with respect to chemical equilibrium is the fact that either of eqn (2.261) no longer gives the law of mass action. Using the Debye–Hückel formula for µ we now investigate what relation results from the concentration at equilibrium for this form of the chemical potential. Consider the reaction

$$HA \rightleftharpoons H^+ + A^-, \quad K, \qquad (2.264)$$

which is a simple dissociation of a proton producing a pair of singly charged ions. The conservation conditions are taken as

$$[H^+] = [A^-] = x,$$
$$[HA] + [A^-] = c_0, \quad (2.265)$$

while the chemical potentials are

$$\mu_{HA} = \mu^0_{HA} + RT\ln[HA]$$
$$\mu_{H^+} = \mu^0_{H^+} + RT\ln[H^+] - \alpha'[H^+]^{\frac{1}{2}},$$
$$\mu_{A^-} = \mu^0_{A^-} + RT\ln[A^-], \quad (2.266)$$
$$\alpha' = \frac{138.8 \times 10^3}{0.304D},$$

where we have chosen to put all the electrostatic source of nonideality in the μ for H^+ (we could have put it all in the μ for A^- or half in each; the point is that our treatment gives $RT\ln\gamma$ for a pair of ions, e.g. Na^+ Cl^-, and hence the total source of nonideality in the Debye–Hückel theory is included in eqn (2.266)). The general condition for equilibrium (eqn (2.32)) for the reaction of eqn (2.264) is

$$\mu_{HA} = \mu_{H^+} + \mu_{A^-}. \quad (2.267)$$

Using the μ of eqn (2.266) we obtain

$$\frac{[H^+][A^-]}{[HA]}\exp(-\alpha[H^+]^{\frac{1}{2}}) = \exp(-\Delta\mu/RT) = K(T),$$
$$\Delta\mu^0 = \mu^0_{H^+} + \mu^0_{A^-} - \mu^0_{HA}, \quad \alpha = \alpha'/RT, \quad (2.268)$$

or using eqn (2.265) (α = 2.35 for a 1–1 electrolyte in water at 298 K)

$$\frac{x^2}{c_0 - x}\exp(-\alpha x^{\frac{1}{2}}) = K, \quad (2.269)$$

which must be solved numerically (or graphically) for x and is to be compared with the standard law of mass action result

$$\frac{[H^+][A^-]}{[HA]} = K$$

or

$$\frac{x^2}{c_0 - x} = K. \tag{2.270}$$

Note that we can predict that $x_{\text{non-ideal}} > x_{\text{ideal}}$ (where x_{ideal} is the solution of the usual equilibrium-constant expression given in eqn (2.270)) since the reaction of eqn (2.264) produces ions of opposite charge which interact with an attractive Coulombic potential. Thus the extra attractive interaction between ions tends to drive the reaction toward the right in the direction of producing more ions (than would be predicted by ignoring long-range electrostatic interactions). For acetic acid with $K = 10^{-4.75}$, Table 2.14 gives x_{ideal} and $x_{\text{non-ideal}}$ (the solutions of eqns (2.270) and (2.269) respectively) as a function of c_0, the total acid concentration. We see that including the effect of long-range electrostatic interactions in the chemical potential (eqn (2.266)), using the Debye–Hückel formulation, does not have a major effect on the equilibrium concentrations (and the Debye–Hückel theory overestimates the effects of electrostatic interactions).

2.3.5. Excluded volume

In the preceding section we examined how the long-range Coulombic interaction between ions leads to deviations in the form of the chemical potential (eqn (2.261)) from the ideal

TABLE 2.14.

Concentration of dissociated protons as a function of the concentration of acetic acid using the law of mass action (eqn (2.270)) and the Debye–Hückel theory (eqn (2.269))

c_0 (mol l^{-1})	$[H^+]$ Law of mass action	$[H^+]$ Debye–Hückel
0.001	0.000125	0.000126
0.01	0.000413	0.000423
0.1	0.00133	0.00138
1.0	0.00421	0.00455

form for independent molecules that gives the law of mass action (eqn (2.87)). In this section we explore another source of non-ideality that gives departures from the law of mass action. But rather than a long-range energy effect as with the Coulombic interaction between ions, the present case is a short-range entropy effect, namely the fact that a molecule occupies space thus excluding part of the total volume of a system from occupancy by othermolecules.

To give an explicit, albeit very simple, example we turn to the lattice model discussed in section 2.1.5. Consider a lattice of M sites (M is the analogue of V for the lattice model) where two species A and A_2 can occupy any vacant lattice sites with the state of binding not influenced by whether or not a neighbouring site is occupied. We define

N_A = number of A molecules on the lattice,

N_{A2} = number of A_2 molecules on the lattice,

q_A = partition function for a single A molecule,

q_{A2} = partition function for a single A_2 molecule.

Let A and A_2 interconvert by the reaction

$$2A \rightleftharpoons A_2, \qquad (2.271)$$

which leads to the conservation relation

$$N_A + 2N_{A2} = N_0, \qquad (2.272)$$

N_0 being the total number of A molecules in any form. Then the partition function is (see section 2.1.5)

$$\begin{pmatrix}\text{multiple occupancy}\\ \text{of a site forbidden}\end{pmatrix} Q = \sum_{N_A} \sum_{N_{A2}} \frac{M!}{(M-N_A-N_{A2})! \, N_A! \, N_{A2}!} q_A^{N_A} q_{A2}^{N_{A2}}. \qquad (2.273)$$

If we were to ignore the fact that sites occupied by A and A_2 molecules are excluded from occupancy by other particles, then Q would be

$$\left(\begin{array}{l}\text{multiple occupancy}\\ \text{of a site not}\\ \text{forbidden}\end{array}\right) \quad Q = \sum_{N_A} \sum_{N_{A_2}} \frac{(Mq_A)^{N_A}}{N_A!} \frac{(Mq_{A2})^{N_{A2}}}{N_{A_2}!} . \qquad (2.274)$$

We have already seen that the form of eqn (2.273) reduces to that of eqn (2.274) as N_A/M and N_{A_2}/M go to zero (eqn (2.93)). For both eqn (2.273) and eqn (2.274) the chemical potentials are given by

$$\mu_A/kT = -\frac{\partial \ln Q}{\partial N_A},$$

$$\mu_{A_2}/kT = -\frac{\partial \ln Q}{\partial N_{A_2}}. \qquad (2.275)$$

The general criterion for equilibrium for the reaction of eqn (2.271) is

$$2\mu_A = \mu_{A_2}. \qquad (2.276)$$

To obtain the derivatives we need use only the general terms of eqn (2.273) and (2.274) and Stirling's approximation for the factorials (this operation together with eqn (2.276) is equivalent to the maximum-term approximation of section 2.1.5). We obtain (with $\vartheta_A = N_A/M$, $\vartheta_{A2} = M_{A2}/M$)

$$\left(\begin{array}{l}\text{ideal — ignoring}\\ \text{excluded volume}\end{array}\right) \quad \begin{array}{l} \mu_A = -kT\ln q_A + kT\ln\vartheta_A, \\ \\ \mu_{A_2} = -kT\ln q_{A_2} + kT\ln\vartheta_{A_2}, \end{array} \qquad (2.277a)$$

$$\left(\begin{array}{l}\text{non-ideal —}\\ \text{excluded volume}\end{array}\right) \quad \begin{array}{l} \mu_A = -kT\ln q_A + kT\ln\vartheta_A - kT\ln(1-\vartheta_A-\vartheta_{A_2}), \\ \\ \mu_{A_2} = -kT\ln q_{A_2} + kT\ln\vartheta_{A_2} - kT\ln(1-\vartheta_A-\vartheta_{A_2}). \end{array} \qquad (2.277b)$$

Note that when multiple occupancy of a site (excluded volume) is included, the chemical potentials of A and A_2 (eqn (2.277b)) are both functions of ϑ_A and ϑ_{A2}, i.e. the chemical potential μ_i is no longer dependent only on the concentration of species

i as in eqn (2.277a). Applying eqn (2.276), the general criteria for chemical equilibrium, we have

(ideal) $$\frac{\vartheta_{A2}}{\vartheta_A^2} = \frac{q_{A2}}{q_A^2} = K, \qquad (2.278a)$$

(non-ideal) $$\frac{\vartheta_{A2}(1-\vartheta_A-\vartheta_{A2})}{\vartheta_A^2} = \frac{q_{A2}}{q_A^2} = K. \qquad (2.278b)$$

Of course as ϑ_A and ϑ_{A2} go to zero, eqn (2.278b) approaches eqn (2.278a). Using eqn (2.272) to eliminate ϑ_{A2} (with $c_0 = N_0/M$), eqn (2.278) becomes

(ideal) $$(1-x) = \alpha x^2 \qquad (2.279a)$$

(non-ideal) $$(1-x)\{1-\tfrac{1}{2}c_0(x+1)\} = \alpha x^2, \qquad (2.279b)$$

$$\alpha = 2Kc_0, \qquad x = \vartheta_A/c_0.$$

Table 2.15 shows x_{ideal} and $x_{non-ideal}$ calculated from eqn (2.279) for various values of c_0 and K. For Table 2.15 to be useful we must be able to convert from the lattice density c_0 to densities (or concentrations) of gases and solutions. If we consider the lattice spacing to be of the same order as a molecular diameter (such that $c = 1$ corresponds to a condensed phase with essentially no vacancies) then

$$c_0 \simeq \frac{\text{concentration of gas or solution}}{\text{concentration of pure phase}}.$$

For example, the concentration of liquid oxygen is ~ 36 mol l^{-1} while the density of an ideal gas at 273 K and 100 kPa pressure (see section 2.3.2) is 0.045 mol l^{-1}. Thus for oxygen gas at 273 K and 100 kPa, $c_0 \simeq 10^{-3}$ which is very small and is consistent with the fact that under these conditions oxygen behaves as an ideal gas. We note that even for solutions as concentrated as 1 M, where $c_0 \sim 0.02$ (for molecules of the size of H_2O), excluded volume effects are small, producing only minor changes from the equilibrium concentrations predicted by the law of mass action.

TABLE 2.15.

The quantity x of eqn (2.279) as a function of K and c_0. The entry in the table gives the value of x calculated taking excluded volume into account, while the value in parenthesis is the result of the law of mass action, excluded volume ignored

c_0 \ K	10.0	1.0	0.1	0.01
1.0	0.137 (0.200)	0.333 (0.500)	0.613 (0.854)	0.833 (0.981)
0.5	0.231 (0.270)	0.535 (0.618)	0.861 (0.916)	0.981 (0.990)
0.1	0.487 (0.500)	0.843 (0.854)	0.979 (0.981)	0.998 (0.998)
0.01	0.853 (0.854)	0.981 (0.981)	0.998 (0.998)	1.000 (1.000)

Of course the lattice model we have just treated is a highly oversimplified model (molecules on specific lattice sites, qs independent of neighbours); it does, however, give the order of magnitude to be expected for excluded volume effects and shows that in general the μ are functions of the concentrations of all species in the system (eqn (2.277b).

The model we have just treated estimates the effects of excluded volume for a reaction occurring in the gas phase or in solution. We now consider the effect of excluded volume on the entropy of a condensed phase using the simplest model possible, a fluid of hard spheres. If d is the diameter of the sphere, the potential energy is zero if $r > d$ and is infinite if $r < d$; hence the internal energy has no potential-energy term. The thermodynamic functions of such a fluid can be written

$$U = \tfrac{3}{2}NkT,$$

$$S = S^0 + Nk\ln(V_{eff}/N)$$
$$= S^0 - Nk\ln\rho + Nk\ln\nu,$$

$$\mu = \mu^0 + kT\ln\rho + kT\ln\gamma,$$

(2.280)

AND INTERMOLECULAR FORCES 157

where (see eqn (2.86))

$$\rho = N/V,$$
$$\nu = V_{eff}/V,$$
$$\mu^0 = kT\ln\Lambda^3,$$
$$S^0 = -Nk\ln\Lambda^3 + \frac{5}{2}Nk.$$
(2.281)

Note that for this system, the factor ν that contributes the non-ideal portion of S is strictly simply the ratio of the total volume V to the effective volume (excluding the space occupied by other molecules). From thermodynamics at constant T and N,

$$dG = Vdp,$$
$$d\mu = dg = \frac{V}{N}dp = \rho^{-1}dp = \rho^{-1}\frac{p}{\rho}d\rho,$$
(2.282)
$$dF = -pdV = -p\frac{N}{\rho^2}d\rho = -TdS.$$

Expressing the equation of state as a series in the density (eqn (2.215))

$$p/\rho kT = 1 + \sum_{n=1}^{\infty} B_{n+1}\rho^n,$$
(2.283)

and using eqn (2.283) in eqn (2.282) we obtain

$$\mu = \mu^0 + kT\ln\rho + kT\sum_{n=1}^{\infty}\left(\frac{n+1}{n}\right)B_{n+1}\rho^n,$$
(2.284)
$$S = S^0 - Nk\ln\rho - Nk\sum_{n=1}^{\infty}\left(\frac{1}{n}\right)B_{n+1}\rho^n.$$

Thus

$$\ln \gamma = \sum_{n=1}^{\infty} \left(\frac{n+1}{n}\right) B_{n+1} \rho^n ,$$

(2.285)

$$\ln \nu = - \sum_{n=1}^{\infty} \left(\frac{1}{n}\right) B_{n+1} \rho^n ,$$

The first coefficient in eqn (2.285) B_2, is the second virial coefficient given by eqn (2.216) ($\varphi(r) = 0$; $\varphi(r) = +\infty$, $r < d$)

$$B_2 = 2\pi \int_0^d r^2 dr = \frac{2}{3}\pi d^3 .$$

(2.286)

The virial coefficients up to B_6 have been calculated numerically. Introducing

$$\rho_0 = \left(\frac{\text{density at}}{\text{close packing}}\right) = \sqrt{2}/d^3 ,$$

(2.287)

and defining

$$\rho_r = \rho/\rho_0 ,$$

$$B'_{n+1} = B_{n+1} \rho_0^n ,$$

(2.288)

eqn (2.285) becomes

$$\ln \gamma = \sum_{n=1}^{\infty} \frac{n+1}{n} B'_{n+1} \rho_r^n ,$$

(2.289)

$$\ln \nu = - \sum_{n=1}^{\infty} \frac{1}{n} B'_{n+1} \rho_r^n .$$

The density ρ_r relative to close packing is a parameter similar to c_0 in the lattice model just discussed. At close packing, the spheres are arranged on a regular lattice (hexagonal close packing); for $\rho_r = 0.5$ half the spheres present at close packing are missing. Of course, unlike the lattice model, for

$\rho < \rho_r$ the particles are not restricted to regular lattice sites. The quantities B'_n to B'_6 are [30]

$$B'_2 = 1.8138 \qquad B'_4 = 3.1787$$
$$B'_3 = 2.5727 \qquad B'_5 = 3.6128. \qquad (2.290)$$
$$B'_6 = 3.9105$$

Using a powerful technique for estimating the complete series in eqn (2.289) based on a limited number of initial terms (eqn (2.290)) known as Pade' approximants, we have [31]

$$\ln \nu = -\rho_r \left[\frac{2.962 - 0.971\rho_r + 0.122\rho_r^2}{1.000 - 1.253\rho_r + 0.362\rho_r^2} \right],$$
$$\qquad (2.291)$$

$$p/\rho kT = 1 + \rho_r \left[\frac{2.962 - 0.555\rho_r + 0.450\rho_r^2}{1.000 - 1.655\rho_r + 0.710\rho_r^2} \right].$$

Eqns (2.291) are derived using B'_2 to B'_6; since $\ln \nu$ and $p/\rho kT$ are given as the ratio of two polynomials, a Taylor-series expansion would give an infinite series in powers of ρ_r, the coefficients of the terms up to ρ_r^5 being identical with the original series. The use of a few terms in a series to estimate the rest of the series is an extremely powerful tool for the analysis of dense systems.

Another approximate theoretical approach yields the relations [32]

$$\ln \nu = \ln(1-\zeta) - \frac{3\zeta(1-\tfrac{1}{2}\zeta)}{(1-\zeta)^2}$$

$$p/\rho kT = \frac{1 + \zeta + \zeta^2}{(1-\zeta)^2}, \qquad (2.292)$$

$$\zeta = \frac{\pi}{3\sqrt{2}} \rho_r.$$

Table 2.16 gives $\ln \nu$ as a function of ρ_r and $\ln \nu$ given by eqn (2.291), eqn (2.292), and by the series of eqn (2.289) as a

TABLE 2.16.

The effect of excluded volume on the entropy of a fluid of hard spheres; $V_{eff} = \exp(-\nu)V$

$-\ln \nu$ ρ_2	eqn (2.285) up to B_2'	eqn (2.285) up to B_4'	eqn (2.285) up to B_6'	eqn (2.291)	eqn (2.291)
0.2	0.592	0.722	0.726	0.726	0.727
0.4	1.185	1.782	1.855	1.855	1.879
0.6	1.777	3.301	3.713	3.843	3.945
0.8	2.370	5.397	6.843	7.904	8.425

function of the number of terms included. The results of eqns (2.291) and (2.292) agree closely and also agree extremely well with computer simulations[33] (these simulations incidentally show that there is apparently a phase transition to an ordered state at $\rho_r \simeq 0.61$). To see the effects of excluded volume on the entropy we write the entropy as

$$S/Nk = S_{ideal}/Nk + \ln \nu. \qquad (2.293)$$

The entropy of ideal gases falls in the range 80–160 J mol^{-1} K^{-1}. Thus S_{ideal}/Nk (per mole this is $S_{ideal}/R = S_{ideal}/8.4$) is in the range 10–20 (S/Nk is dimensionless). The quantity $\ln \nu$ thus is the amount by which S/Nk is reduced from the value S_{ideal}/Nk. We see that relative to $S_{ideal}/Nk = 10$–20, $\ln \nu$ does not represent a very large reduction in the entropy of the system except for very high densities. Since (see eqn (2.280)) $\nu = V_{eff}/V$ is the ratio of the actual volume available to a molecule to the total volume, Table 2.16 gives $\nu = \exp(-1.23)$ at $\rho_r = 0.3$ and $\nu = \exp(-3.95)$ at $\rho_r = 0.6$, giving an estimate of the reduction in available volume as a function of density.

2.3.6. Non-ideality from association

In the last two sections we have discussed two sources of non-ideality (deviation from the law of mass action), one an energy effect (long-range Coulombic interaction between ions) and the other an entropic effect (the finite size of molecules reducing the available volume). For solute concentrations less

AND INTERMOLECULAR FORCES 161

than 1 M, these effects cause deviations in equilibrium concentrations that at most differ about 25 per cent from those predicted by the law of mass action (the error of course becoming less as the solutions become more dilute). In this section we want to discuss a source of non-ideality that can often alter the results of the law of mass action by factors of 10. The non-ideality we are discussing is when one or all of the components taking part in a reaction also take part in other reactions such as self aggregation.

For example, suppose that the reaction of interest is the complexing of molecule A with molecule B

$$A + B \rightleftharpoons A\text{-}B, \qquad K_{AB}. \qquad (2.294)$$

Applying the law of mass action and the appropriate conservation relations, the concentration of A, B, and AB at equilibrium can be calculated from the equations

$$\frac{[A\text{-}B]}{[A][B]} = K_{AB},$$

$$[A] + [A\text{-}B] = A_0, \qquad (2.295)$$

$$[B] + [A\text{-}B] = B_0.$$

Now suppose that self-aggregates are given by eqn (1.104). Proceeding as in section 2.1.8, the grand partition function for this system is

$$\Xi = \exp(Vq_B\lambda_B) \exp(Vq_{AB}\lambda_A\lambda_B) \exp(\sum_n Vq_n\lambda_A^n), \qquad (2.296)$$

where q_B, q_{AB}, and q_n are the molecule partition functions for the respective species ($K_{AB} = q_{AB}/q_1 q_B$). As usual we find λ_A and λ_B from the relations

$$N_A^* = \frac{\partial \ln \Xi}{\partial \ln \lambda_A} = Vq_{AB}\lambda_A\lambda_B + V\sum_n nq_n\lambda_A^n,$$

$$(2.297)$$

$$N_B^* = \frac{\partial \ln \Xi}{\partial \ln \lambda_B} = Vq_B\lambda_B + Vq_{AB}\lambda_A\lambda_B,$$

or

$$q_{AB}\lambda_A\lambda_B + \sum_n nq_n\lambda_A^n = A_0$$

$$q_B\lambda_B + q_{AB}\lambda_A\lambda_B = B_0 \tag{2.298}$$

Eliminating λ_B one has a relation for λ_A:

$$\lambda_B = B_0/(q_B + q_{AB}\lambda_A),$$

$$\frac{q_{AB}\lambda_A B_0}{(q_B + q_{AB}\lambda_A)} + \sum_n nq_n\lambda_A^n = A_0. \tag{2.299}$$

For arbitrary q_n it is impossible to obtain an explicit solution of eqn (2.299) for λ_A although we could obtain numerically the value of λ_A that satisfies eqn (2.299) if the q_n are known (or have been modelled). It is clear from eqn (2.299) that λ_A and λ_B are functions both of A_0 and B_0 (recall that in general $\mu_A = kT\ln \lambda_A$ and $\mu_B = kT\ln \lambda_B$). Making the obvious interpretation of eqn (2.298) as

$$[A\text{-}B] + \sum_n n[A_n] = A_0,$$

$$[B] + [A\text{-}B] = B_0, \tag{2.300}$$

then we have

$$\lambda_A = [A]/q_1,$$

$$\lambda_B = [B]/q_B. \tag{2.301}$$

Utilizing the dimensionless quantities

$$\beta = A_0 \frac{q_{AB}}{q_1 q_B} = A_0 K_{AB}, \qquad \rho = B_0/A_0,$$

$$\alpha_n = A_0^{n-1} \frac{q_n}{q_1^n} = A_0^{n-1} K_n, \qquad x = [A]/A_0, \tag{2.302}$$

eqn (2.299) becomes

$$1 - x(1-\beta+\beta\rho) - \beta x^2 - (1-\beta x)\sum_{n=2}^{\infty} n a_n x^n = 0, \qquad (2.303)$$

(with self-aggregation of A)

which is to be compared with the case where there is no self-aggregation of A (eqn (2.295) using the definitions of eqn (2.302)):

$$1 - x(1-\beta+\beta\rho) - \beta x^2 = 0 \text{ (no self-aggregation of A)}. \quad (2.304)$$

In this chapter we have discussed three specific examples of deviations from the law of mass action: eqn (2.269) (deviations because of long-range Coulombic interactions); eqn (2.279b) (deviations because of excluded volume); and eqn (2.303) (deviations because of self-aggregation of one of the components). The source of deviation from the law of mass action when other reactions (such as self-aggregation) are present is qualitatively different from the other sources of nonideality we have discussed: we have used the law of mass action (see the development following eqn (1.104)) to obtain deviations from the law of mass action. Strictly the foregoing statement does not make sense. We mean it to be taken in the following practical manner. If we consider a single reaction like eqn (2.294), then the application of eqn (2.295) (or equivalently eqn (2.304)) often gives very inaccurate equilibrium concentrations if other reactions are ignored. For example, the application of the law of mass action to the binding of urea to a protein will be very inaccurate if we ignore the self-association of urea. If we include all significant reactions, treating each by the law of mass action (which means we treat each possible species as ideal, which in turn means that each has a molecular partition function of the form (V_{qi}) and that excluded volume is ignored) then in general we can obtain excellent results. My point is that the isolated application of eqn (2.295), for example, is often incorrect. To treat many coupled reactions we emphasize that the only systematic treatment is via the grand partition function (which is also the simplest treatment).

2.3.7. Thermodynamics of argon (solid-liquid-gas)

In this section we apply the concepts and relations discussed in previous sections to the understanding of the thermodynamic functions of argon. My purpose is not to give the most sophisticated treatment possible, but rather to give the simplest interpretation that accurately reflects the major factors contributing to the thermodynamic functions in the various phases of argon with reference to the role played by intermolecular forces in each phase.[34]

Table 2.17 lists the values of the thermodynamic functions H, S, G, and C_p for argon as a function of temperature at the constant pressure of one atmosphere.[35]

We start our survey of the data in Table 2.17 at 0 K and examine the sublimation energy. From X-ray studies the distance of closest approach of two argon atoms in the crystal at 0 K is $R_m = 0.372$ nm. Since argon packs like hard spheres with twelve nearest neighbours, we can use ΔU_{subl} (which is $-H(0\ K)$) and R_m in eqn (2.236) to construct an inter-atom pair potential. Choosing $t = 12$ and $s = 6$ for simplicity, eqn (2.236) gives

$$\epsilon = 7783/8.7 = 894\ \text{J mol}^{-1},$$
$$r_m = 0.372/0.97 = 0.384\ \text{nm}. \tag{2.305}$$

With the '6–12' potential, at r_m the attractive van der Waals energy between a pair of argon atoms is $-2\epsilon = -1.789$ kJ mol^{-1} (this compares favourably with $-C_6/r_m^6$ calculated from London's equation, eqn (2.205)), there being $\epsilon = 0.894$ kJ mol^{-1} of repulsive energy yielding a net energy of $-\epsilon$. The potential energy of the crystal per particle or mole of particles is approximately -6ϵ, eqn (2.234) giving a better estimate (there are twelve nearest neighbours but each pair energy $-\epsilon$ is shared by two atoms, giving an energy per atom of -6ϵ due to nearest neighbours; the factors M_t and M_s in eqn (2.263) include interactions beyond nearest neighbours with all the atoms in the crystal). There is a small amount of kinetic energy at 0 K due to zero-point vibration, but we ignore that correction here. Thus the thermodynamic functions at 0 K are

TABLE 2.17.

Thermodynamic data for argon (pressure = 100 kPa)

T (K)	State	H (J mol^{-1})	S (J K^{-1} mol^{-1})	G (J mol^{-1})	C_p (J K^{-1} mol^{-1})
0	solid	−7783	0.00	−7783	0.00
20	solid	−7686	6.30	−7812	11.84
40	solid	−7308	18.61	−8051	22.26
60	solid	−6846	28.48	−8560	26.80
80	solid	−6258	36.96	−9215	32.34
83.8	solid	−6090	38.47	−9311	33.60
83.8	liquid	−4872	53.34	−9311	42.42
87.3	liquid	−4830	55.02	−9618	46.62
87.3	gas	+1806	131.04	−9618	21.00

Melting point = 83.8 K
Boiling point = 87.3 K
$\Delta H_{subl} = \Delta U_{subl} = 7783$ J mol^{-1}
$\Delta H_{fusion} = +1218$ J mol^{-1}
$\Delta S_{fusion} = 14.87$ J K^{-1} mol^{-1}
$\Delta H_{vaporization} = +6636$ J mol^{-1}
$\Delta S_{vaporization} = 76.02$ J K^{-1} mol^{-1}
Triple point at 83.85 K, $p = 67$ kPa
Critical point at 151.15 K, $p = 4864$ kPa

$$G = F = U = H = \Phi = -\Delta U_{subl},$$
$$S = 0, \qquad (2.306)$$
$$C_V = C_p = 0,$$

where Φ is the crystal potential energy ($\sim -6\varepsilon$) given by eqn (2.234) with $R = R_m = t = 12$, $s = 6$).

As the temperature is increased, the vibrational levels of the crystal are progressively populated. As the 'vibrational ladder' is climbed the potential energy becomes more positive with a concomitant increase in kinetic energy (which is positive). The entropy increases since the number of accessible states increases with temperature. Since both the potential energy and kinetic energy are changing (both becoming more positive), C_V and C_p, which measure the change in energy with temperature, increase. From Table 2.18 we see that the density (or volume/atom) does not change significantly with temperature. Assuming that the volume is constant, $H = U$, $G = F$, and $C_p = C_V$. Using the Einstain model to calculate the effect of vibrational excitation on the thermodynamic functions, we obtain the results of Table 2.19 using ϑ_{vib} = 80 K in eqn (2.75) (in principle ϑ_{vib} can be calculated from the inter-atom pair potential $\varphi(r)$, but here it is taken as an empirical parameter).

TABLE 2.18.

Density of argon and average distance between atoms as a function of temperature (pressure = 100 kPa) (the distance of closest approach between two atoms in the crystal at 0 K is 0.372 nm)

T (K)	State	Density (kg m^{-3})	Volume/molecule (nm^3)	$\langle R \rangle$ (nm)
20.4	solid	1762	0.0377	0.374
40.8	solid	1735	0.0384	0.378
60.6	solid	1693	0.0392	0.381
79.5	solid	1638	0.0406	0.384
83.8	solid	1622	0.0410	0.386
87.3	liquid	1402	0.0475	0.405
87.3	gas	61	10.9000	2.500
151.15	vapour[†]	531	0.1250	0.560

[†]At the critical temperature with p_c = 4900 kPa.

TABLE 2.19.

Thermodynamic functions for the Einstein model of solid argon (ϑ_{vib} = 80 K) (units are the same as in Table 2.17)

T (K)	U (=H)	U-U (0 K)	S	F (=G)	C_V (=C_p)
0	-7783	0	0.00	-7783	0.00
20	-7762	21	11.34	-7801	7.56
40	-7472	1554	57.96	-7934	18.14
60	-7090	3465	98.28	-8274	21.67
80	-6615	5838	130.83	-8698	23.18

Comparing the experimental values of the thermodynamic functions for the solid phase in Table 2.17 with the values given by the Einstein model in Table 2.19, we see that the trends and magnitudes are well accounted for. In particular note that U does not change markedly from its value at 0 K (U (0 K) = $\Delta U_{subl} = \Phi$), as is indicated in the column ($U-U$ (0 K)). At 80 K the crystal has, however, achieved a sizeable entropy (26.2 J mol^{-1} K^{-1} in the Einstein model, 37 J mol^{-1} K^{-1} experimentally) which is to be compared with the entropy that is eventually achieved in the gas phase ($\sim$ 120 J mol^{-1} K^{-1}). Thus at the melting point the crystal has almost one-third of the entropy it will have in the gas phase at the boiling point.

We now move on to the gas phase. Using the model of an ideal gas, eqn (2.86) gives (per mole at 87.3 K)

$$H = U + pV = \frac{3}{2} RT + RT = \frac{5}{2} RT = 1822 \text{ J mol}^{-1},$$

$$C_p = \left(\frac{\partial H}{\partial T}\right)_P = \frac{5}{2} R = 521 \text{ J K}^{-1} \text{ mol}^{-1}. \quad (2.307)$$

The values obtained in eqn (2.307) are seen to be in excellent agreement with the experimental values given in Table 2.17. It is instructive to compare the internal energy of the gas at 87.3 K ($\frac{3}{2} RT$) with that of the crystal at 0 K. We have

$$U = E_{pot} + E_{kin},$$

$\begin{pmatrix}\text{crystal}\\\text{at 0 K}\end{pmatrix}$ $\quad U = -7783 + \sim 0 = 7783 \text{ J mol}^{-1},\quad (2.308)$

$\begin{pmatrix}\text{gas at}\\\text{87 K}\end{pmatrix}$ $\quad U = \sim 0 \quad + 1100 = + 1100 \text{ J mol}^{-1}.$

Thus the internal energy goes from being negative and all potential energy in the crystal at 0 K to being positive and all kinetic energy in the gas at 87 K.

To calculate the entropy we use the Sackur–Tetrode equation (eqn (2.88) with molecular weight = 40 for argon)

$$S = 57.43 + 28.84 \log T - 17.97 \log c \quad (\text{J K}^{-1} \text{ mol}^{-1}) \quad (2.309)$$

where c is the concentration in mol l^{-1}. Since the conditions are constant pressure (100 kPa) we have for the temperature dependence of c

$$pV = nRT,$$

$$\frac{c(273 \text{ K})}{c(T)} = \frac{T}{273},$$

$$c(273 \text{ K}) = 1/22.4,$$

$$c(T) = 12.2/T.$$

(2.310)

Using the last of eqn (2.310), eqn (2.309) becomes

$$S = 33.86 + 46.82 \ln T, \qquad (2.311)$$

giving at 87.3 K

$$S = 126 \text{ J K}^{-1} \text{ mol}^{-1}, \qquad (2.312)$$

which is seen to be in excellent agreement with the value given in Table 2.17.

We turn now to a consideration of the liquid phase. The liquid phase is the most difficult phase to treat since one can assume neither that it has a simple geometric arrangement of atoms (as in the crystal) nor that the atoms are independent (as in the gas). However, we can understand the magnitude of the thermodynamic functions in a simple manner. From Table 2.19 we observe that C_V at 80 K has almost reached the classical value of the law of Dulong and Petit of $C_V = 3R = 25$ J K^{-1} mol^{-1} (see the paragraph following eqn (2.75)). Thus to a very good approximation one can treat the liquid using classical statistical mechanics. From eqn (2.167) one has for the liquid

$$U = \frac{3}{2}RT + \langle E_{pot} \rangle. \qquad (2.313)$$

Since at 100 kPa pressure argon exists in the liquid phase only over a temperature range of about 5 K, we need only examine the thermodynamic functions at a single temperature (at 100 kPa pressure we are only slightly above the triple point; for

$p < 66$ kPa the crystal sublimes directly to the gas phase, there being no liquid phase). Using the experimental value of U at 84 K from Table 2.17 we can use eqn (2.313) to calculate the average potential energy

$$\langle E_{pot} \rangle = U - \tfrac{3}{2}RT = -4849 - 1066 = -5915 \text{ J mol}^{-1}. \quad (2.314)$$

Comparing the average potential energy of the liquid, -5915 J mol^{-1}, with the potential energy of the crystal at 0 K, -7783 J mol^{-1}, we see that the liquid has 76 per cent of the potential energy of the crystal. Using the density of the liquid given in Table 2.18 we can use the average distance between atoms (assuming hexagonal packing) to estimate the average potential energy if the liquid were arranged on a perfect hexagonal lattice with nearest-neighbour distances given by the density of the liquid. Since this introduces much more order than the liquid actually has, the potential energy so estimated will be an overestimate (more negative). First note that $\langle R \rangle = 0.405$ nm is only about 0.025 nm larger than the nearest-neighbour distance in the crystal. Using eqn (2.234) with $t = 12$, $s^* = 6$, $R_m = 0.372$ nm, $R = \langle R \rangle = 0.405$ nm, $\Delta U_{subl} = +7783$, we have

$$\Phi(0.405) = 7783\left\{\left(\frac{0.372}{0.405}\right)^{12} - 2\left(\frac{0.372}{0.405}\right)^{6}\right\} = -6521 \text{ J mol}^{-1}. \quad (2.315)$$

Thus the view of the liquid as a slightly expanded crystal overestimates the average potential energy of the liquid, but comparing the result of eqn (2.315) with that of eqn (2.314) we see that the estimate is quite good. We are of course assuming in eqn (2.314) that $U \simeq H$ for a condensed phase which is quite a good approximation (though not for the gas phase — see eqn (2.307)).

The entropy of the liquid, ~ 55 J mol^{-1} K^{-1}, is interesting in that it is about half way between the entropy of the crystal at 0 K (0 J mol^{-1} K^{-1}) and that of the gas at 87 K (~ 126 J mol^{-1} K^{-1}). We can gain some understanding of the entropy of the liquid by applying eqn (2.288) in the form (evaluating the constants for $T = 84$ K and molecular weight $= 40$)

$$S = 113.03 - 17.97 \log c_{eff}. \quad (2.316)$$

Treating the effective concentration c_{eff} as a parameter that we can calculate using the experimental value of S from Table 2.17 we obtain

$$c_{eff} = 10^{3.32} \text{ mol l}^{-1}. \quad (2.317)$$

This corresponds to an effective volume per atom of

$$\text{volume/atom} = 1.26 \times 10^{-3} \text{ nm}^3/\text{atom} \quad (2.318)$$

Interpreting the volume per atom as the effective sphere in which the point centre of an atom can translate, we have that the radius of this sphere is given by

$$\text{volume/atom} = \tfrac{4}{3}\pi R^3 = 1.26 \times 10^{-3} \text{ nm}^3/\text{atom} \quad (2.319)$$

$$R \simeq 0.067 \text{ nm}.$$

Note that we can estimate an effective or free volume per atom from the data of Table 2.18. Thus

$$V_{eff}/\text{atom} = V_{free}/\text{atom} = V(84 \text{ K}) - V(0 \text{ K}) \simeq$$
$$\simeq 0.047 - 0.037 = 0.01 \text{ nm}^3/\text{atom}. \quad (2.320)$$

The effective volume/atom estimated from eqn (2.320) is equivalent to a concentration of 166 mol l^{-1}. Using this value in eqn (2.316) gives an entropy of 74 J mol^{-1} K^{-1}, which is higher than the experimental value of 55 J mol^{-1} K^{-1} but gives the order of magnitude quite well.

My point in the above discussion is that the fact that the entropy of liquid argon is about half that of gaseous argon can be simply rationalized in terms of estimates of the effective volume per atom. The accurate calculation of the entropy of a liquid from first principles remains a challenge to present research in statistical mechanics. Note in particular that we have used the experimental values of the density to estimate the thermodynamic functions; an *a priori* treatment would calculate (predict) the density.

Since at 100 kPa pressure, the liquid phase makes only a brief appearance in the range of 84-7 K (below 66 kPa it does not appear at all), the major phases of argon are the crystal and the gas. The equilibrium

$$\text{solid} \rightleftharpoons \text{gas} \tag{2.321}$$

can be understood in very simple terms taking

$$G_{solid} \sim F_{solid} \sim U(0\ K) = -7783\ J\ mol^{-1},$$

$$G_{gas} = H_{gas} - TS_{gas} = -12.5\ T - 46.8\ T\ \log T, \tag{2.322}$$

where we have used eqn (2.307) and eqn (2.311) to give G_{gas}, and we make the approximation that the crystal is dominated by the potential energy and take the free energy as constant, equal to the potential energy at 0 K. Since for a pure phase the chemical potential is the Gibbs free energy per mole (eqn (2.40)), the temperature at which the gas will be in equilibrium with the crystal (phase transition temperature) is given by

$$G_{gas}(T) = G_{solid}(T). \tag{2.323}$$

Note that eqn (2.323) is an exact result if the exact G_{gas} and G_{solid} are used. Using the approximations of eqn (2.322), Fig. 2.9 shows G_{gas} and G_{solid} as a function of temperature. The temperature at which the two functions cross is the temperature of the phase transition (here ignoring the short temperature range of the liquid). Below the transition temperature, the solid has the lower free energy (being approximately all potential energy); as the temperature is increased the $-TS$ term in the free energy of the gas becomes more and more negative, and at the transition temperature is as large (negative) as the potential energy of the solid; for the temperature greater than the transition temperature, the free energy of the gas is lower than that of the solid. To a good first approximation the solid-gas phase transition can be viewed as an energy-entropy competition

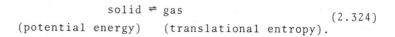

$$\text{solid} \rightleftharpoons \text{gas} \qquad (2.324)$$
$$\text{(potential energy)} \quad \text{(translational entropy)}.$$

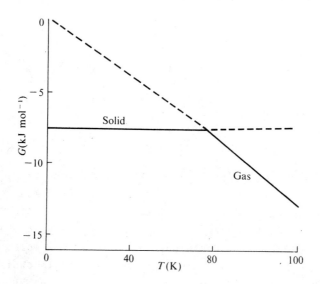

FIG. 2.9. Plot of the free energy of argon at constant pressure for the gas and solid phases. The curves were calculated assuming an ideal gas and a perfect solid described solely by the potential energy at 0 K. The solid curve is for the state of lowest free energy at a given temperature. In this simple picture a phase transition from the low-energy solid to the high-entropy gas is predicted to occur at about 80 K.

We conclude our discussion of the properties of argon by returning to Table 2.18, which lists the average distance between atoms as obtained from the density (eqns (2.243) and (2.245)). The average potential energy and at least an estimate of the entropy, as we have seen from our discussion of the liquid, are readily obtained from a knowledge of $\langle R \rangle$. Thus $\langle R \rangle$

changes very little as the crystal is heated from 0 K to 84 K; the $\langle R \rangle$ of the liquid is only ~ 5 per cent larger than R_m of the crystal at 0 K; the $\langle R \rangle$ of the gas, 2.5 nm, is large enough that clearly the average potential energy is approximately zero, but note that 2.5 nm is only about 6½ times larger than the diameter of an argon atom; and, at the critical point, the temperature and pressure above which there is no longer a discontinuity in the thermodynamic functions between liquid and gas, the $\langle R \rangle$, 0.56 nm, is only a factor of 1½ times larger than R_m of the crystal at 0 K.

2.3.8. Thermodynamics of water (liquid)

Since water is a far more complicated molecule than argon, my remarks on the thermodynamic functions of this substance of necessity will be qualitative and brief. Table 2.20 lists the thermodynamic functions of water with emphasis on the liquid state.[36] Since in the solid each water molecule is hydrogen-bonded to four other water molecules (tetrahedral coordination) a first approximation to the sublimation energy is

$$-\Delta U_{subl} = \tfrac{1}{2}(4\ \varepsilon_{H-bond}) = -47.2\ \text{kJ mol}^{-1} \qquad (2.335)$$

or

$$\varepsilon_{H-bond} = -29.5\ \text{kJ mol}^{-1}. \qquad (2.336)$$

Note that eqn (2.335) attributes all of the energy at 0 K to hydrogen bonds between an H_2O molecule and its four nearest neighbours and neglects the long-range electrostatic interaction (mainly dipole—dipole interaction) with all the molecules in the crystal past the first coordination shell. Incidentally if the charge distribution of H_2O were approximated by a point dipole, the most stable crystal would have a coordination of twelve nearest neighbours (hexagonal close packing); this in fact is the crystal structure of H_2S. The simplest charge distribution that correctly predicts tetrahedral coordination is a tetrahedral arrangement of point charges, identical positive charges centred on the hydrogen atoms, and identical negative charges placed in the direction of the two lone electron

TABLE 2.20.

Thermodynamic data for water

T (K)	State	H (J mol^{-1})	S (J K^{-1} mol^{-1})
0	solid	−47 460	0.00
20	solid	−47 447	0.34
60	solid	−47 208	6.31
100	solid	−46 691	12.79
140	solid	−45 948	18.93
180	solid	−45 003	24.87
220	solid	−43 869	30.60
260	solid	−42 546	36.16
273.15	solid	−42 084	38.04
273.15	liquid	−36 036	59.77
280	liquid	−35 532	61.86
300	liquid	−33 978	66.88
320	liquid	−32 466	71.90
340	liquid	−30 912	76.49
360	liquid	−29 400	80.67
373.15	liquid	−28 434	83.60
373.15	gas	+12 600	193.12

[†]Values of the entropy relative to the residual entropy of ice at 0 K, 3.43 ≃ $R\ln(3/2)$ J K^{-1} mol.

pairs at a distance from the oxygen nucleus that is the same as the O–H bond length. Water is an example where atom-centred charges only (e.g. positive charges centred on the hydrogen atoms, negative charge centred on the oxygen) are not sufficient to explain the qualitative features of the interaction between molecules (tendency to tetrahedral coordination).

As with liquid argon, we see that the internal energy of the liquid ($U \simeq H$) is not significantly larger than that of the crystal at 0 K. Treating the liquid classically (from the equipartition of energy, eqn (2.168), each rotational degree of freedom contributes $\frac{1}{2}RT$ to the internal energy, three rotational degrees of freedom giving $\frac{3}{2}RT$),

$H \simeq U =$ kinetic energy of translation + kinetic energy of rotation + $\langle E_{pot} \rangle$,

$$U = \frac{3}{2}RT + \frac{3}{2}RT + \langle E_{pot} \rangle. \tag{2.337}$$

Table 2.21 gives $\langle E_{pot} \rangle$ as a function of T using eqn (2.337) and the experimental values of $H(U)$ from Table 2.20. Approximating the potential energy as

$$\langle E_{pot} \rangle = X(T) [-2\epsilon_{H-bond}], \tag{2.338}$$

where $-2\epsilon_{H-bond}$ is the energy per molecule (per mole) in the crystal at 0 K and $X(T)$ is the fraction of hydrogen bonds as a function of temperature in the liquid. Table 2.20 also gives the values of $X(T)$ that fit the temperature dependence of $\langle E_{pot} \rangle$. We see that in this interpretation liquid water has almost the same degree of hydrogen bonding as the solid; the small fraction of broken hydrogen bonds leads to some molecules with coordination higher than 4, leading to the unique feature of water that the liquid is more dense than the solid (tetrahedral coordination leads to a very open crystal lattice). It is to be emphasized that the above discussion is a simple approximate interpretation of the thermodynamic data for water.

A precise formulation of the partition function for water, even using the simple picture of hydrogen bonds and broken

TABLE 2.21.

Potential and kinetic energy of liquid water. The potential energy is to be compared with the value of that for the crystal at 0 K, -47.23 kJ mol^{-1}. X(T) is given in eqn (2.338) (energies in kJ mol^{-1})

T (K)	$\langle E_{kin} \rangle$	$\langle E_{pot} \rangle$	$X(T)$
273.15	6.86	-42.72	0.903
280	7.02	-42.38	0.900
300	7.52	-41.34	0.875
320	8.03	-40.34	0.854
340	8.53	-39.29	0.832
360	9.03	-38.29	0.810
373.15	9.36	-37.66	0.796

hydrogen bonds, is very difficult. The problem is that we must formulate the number of ways we can arrange N water molecules with M hydrogen bonds, then sum over all values of M. While liquid water does retain a large degree of tetrahedral coordination, the molecules are not arranged on a perfectly regular lattice. We can easily visualize a sponge-like labyrinth of bonded molecules with no clear-cut distinction between 'bonded' and 'unbonded' molecules. There have been several attempts to treat liquid water as an equilibrium between more-or-less spherical clusters of ice-like, hydrogen-bonded molecules, introducing empirical parameters to describe the basic energies.[37] More recently computer simulations[38] have been used based on an empirical intermolecular potential ('6–12' or '6-exp' potential plus a tetrahedral arrangement of monopoles). Needless to say, the last word has not been said on the exact nature of this most interesting and important liquid.

2.3.9. Dimerizations

For the general chemical equilibrium

$$a\text{A} + b\text{B} + \ldots \rightleftharpoons c\text{C} + d\text{D} + \ldots \tag{2.339}$$

the equilibrium constant is given by

$$K = \frac{q_C^c q_D^d \ldots}{q_A^a q_B^b \ldots}. \tag{2.340}$$

In the gas phase the partition function per molecule is given by

$$q = q_{tr} q_{rot} q_{vib} q_{elec}. \tag{2.341}$$

The subscripts refer respectively to translation, rotation, vibration, and electronic degrees of freedom. To first approximation the quantities below dominate K,

$$\Delta f_{trans} = cf_{trans}(C) + df_{trans}(D) + \ldots$$
$$- af_{trans}(A) - bf_{trans}(B) - \ldots \quad (2.342)$$

$$f_{trans} = -kT\ln q_{trans},$$

$$\Delta f_{elect} = c\varepsilon_C + d\varepsilon_D + \ldots - a\varepsilon_A - b\varepsilon_B - \ldots,$$

where ε_i is the electronic ground state (or bond energy) of the ith species (excited electronic states make essentially no contribution, the spacing between electronic levels being of the order of an electronvolt or 96 kJ mol^{-1}).

For simple dimerizations

$$A + A \rightleftharpoons A_2, \quad K, \quad (2.343)$$

we can estimate the order of magnitude of K by the expression

$$K \simeq \exp(-\varepsilon/RT)\exp(-\Delta f_{trans}/RT), \quad (2.344)$$

where ε is the bond energy of A_2. The quantity f_{trans} is given by

$$f_{trans} = \tfrac{3}{2}RT - TS_{trans}. \quad (2.345)$$

From eqn (2.312), s_{trans} is of the order of 126 J mol^{-1} K^{-1} for argon (recall that s_{trans} depends on the logarithm of the mass of the molecule and the temperature, that is s_{trans} is relatively insensitive to molecular weight and temperature). At room temperature

$$f_{trans} \simeq 4 \text{ kJ mol}^{-1} - 42 \text{ kJ mol}^{-1} \simeq -38 \text{ kJ mol}^{-1}. \quad (2.346)$$

Thus f_{trans} is dominated by s_{trans}. Taking s_{trans} as approximately constant we have

$$\Delta s_{trans} = s_{trans}(A_2) - 2s_{trans}(A) \simeq -s_{trans} \simeq -126 \text{ J mol}^{-1} \text{ K}^{-1}.$$
$$(2.347)$$

Eqn (2.344) thus becomes

$$K \simeq \exp(-\Delta f/RT),$$
$$\Delta f \simeq \varepsilon + Ts_{trans},$$
(2.348)

or at room temperature

$$\Delta f \simeq \varepsilon + 42 \text{ kJ mol}^{-1}.$$
(2.349)

For dimerization to be favoured ε must be more negative than about -42 kJ mol^{-1}. The following tabulation gives approximate values of ε and the resulting Ks for various dimerizations in the gas phase at room temperature (at 300 K, $RT = 25.2$ kJ mol^{-1}).

Dimer	ε (kJ mol^{-1})	Δf (kJ mol^{-1})	K
Ar–Ar	-1.26	$+40.60$	$10^{-16.1}$
H$_2$O–H$_2$O	-20.9	$+20.9$	$10^{-8.3}$
CH$_3$–C(=O...HO)(OH...O=)C–CH$_3$	-58.60	-16.71	$10^{6.7}$
Na$^+$–Cl$^-$	-381	-334	10^{135}

Thus just from a knowledge of the bond energy and the order of magnitude of the translational entropy we understand that argon is little associated in the gas phase, while acetic acid has a strong tendency to form dimers. The large K for the association of Na$^+$ and Cl$^-$ is misleading to the extent that the Ks for forming larger aggregates are even bigger, the most favourable state being the infinite aggregate, the solid.

2.4. Summary

In this chapter I have presented the notions of thermodynamics, statistical mechanics, and intermolecular forces that are important for an understanding of the construction of models for multiple equilibria in biological systems. While much that we have discussed does not have direct or obvious applications to biological systems (e.g. the thermodynamics of argon) the understanding of the interrelation of thermodynamics, statistical mechanics, and intermolecular forces in a simple case (for example the thermodynamics of argon) is essential to the construction of realistic and useful models.

Notes and References

1. For a treatment of thermodynamics free of intuitive borrowings from statistical mechanics one cannot do better than: Pippard, A.B., *The elements of classical thermodynamics*, Cambridge University Press (1957); available in paperback.
2. Our use of stoichiometric coefficients and a progress variable follows the usage found in: Prigogine, I. and Defay, R., *Chemical thermodynamics*, Longsmans, Green and Co., London (1954).
3. A very readable treatment of quantum mechanics at an intermediate level is: Karplus, M. and Porter, R.N., *Atoms and molecules*, Benjamin, New York (1970).
4. A general introduction to statistical mechanics on an introductory level with many applications is: Hill, T.L., *An introduction to statistical thermodynamics*, Addison-Wesley, Reading, Mass. (1960). For many of the same topics one can find more detail in: Hill, T.L., *Statistical mechanics*, McGraw-Hill, New York (1956). A general reference to intermolecular forces is: Hirschfelder, J.O., Curtiss, C.F., and Bird, R.B., *Molecular theory of gases and liquids*, Wiley, New York (1954).
5. The form of eqn (2.107) can be obtained for the model of eqn (2.100) by expanding $\ln t(N,N_2)$ in a Taylor series in N_2 about $N_2 = N_2^*$.
6. The first entry in ref. 4 contains many illustrations of the use of the grand partition function.
7. See ref. 6 in Chapter 1.
8. McClellan, A.L., *Tables of experimental dipole moments*, Freeman, San Francisco (1963).
9. Dzidic, I. and Kebarle, P., *J. phys. Chem.* **74**, 1466 (1970); Arshadii, M., Yamdagni, R., and Kebarle, P., *J. phys. Chem.* **74**, 1475 (1970).
10. Fredenhagen, K., *Z. anorg. Chem.* **218**, 161 (1934).
11. Bauer, S.H., Beach, J.V., and Simond, J.H., *J. Am. chem. Soc.* **61**, 19 (1939).
12. See, for example: Panofsky, W.K.H. and Philips, M., *Electricity and magnetism*, Addison-Wesley, Reading, Mass. (1955).
13. Poland, D. and Scheraga, H.A., *Biochemistry* **6**, 3791 (1967). The dipole moments (direction and magnitude) for methylformate and formamide are given respectively in: Curl, R.F. Jr., *J. chem. Phys.* **30**, 1529 (1959); Kurland, R.J. and Wilson, E.B., *J. chem. Phys.* **27**, 585 (1957).
14. London, F., *Trans. Faraday Soc.* **33**, 8 (1937).
15. Wilson, E.B. Jr., *Adv. chem. Phys.* **2** (1959).
16. For a review of experimental data see: Pimentel, G.C. and McClellan, A.L., *The hydrogen bond*, Freeman, San Francisco (1960).
17. See the first entry in ref. 13.
18. See, for example: Scheraga, H.A., *Adv. phys. org. Chem.* **6**, 103 (1968); contains many references.
19. For a consideration of bond-angle bending in polypeptides see: Gibson, K.D. and Scheraga, H.A., *Biopolymers* **4**, 709 (1966). For a treatment of

the cyclic alkanes see: Bixon, M. and Lifson, S., *Tetrahedron* **23**, 769 (1967).
20. Kirkwood, J.G. and Westheimer, F.H., *J. chem. Phys.* **6**, 506 (1938).
21. Ben-Naim, A., *J. phys. Chem.* **69**, 1922 (1965). Also see: Frank, H.S. and Evans, M.W., *J. chem. Phys.* **13**, 507 (1945).
22. Nemethy, G. and Scheraga, H.A., *J. chem. Phys.* **66**, 1773 (1962).
23. The Landolt–Börnstein tables contain a great deal of thermodynamic data and parameters (rotational and vibrational temperatures, etc.). Springer-Verlag, Berlin.
24. Discussions of the Madelung constants for ionic solids are found in most physical chemistry texts and in any text on solid-state physics. See, for example: Kittel, C., *Introduction to solid state physics*, Wiley, New York (1961).
25. Kihara, T. and Koba, S., *J. phys. Soc., Japan* **7**, 348 (1952).
26. Momany, F.A., Vanderkooi, G., and Scheraga, H.A., *Proc. natn. Acad. Sci. U.S.A.* **61**, 429 (1968). And: McGuire et al., *Macromolecules* **4**, 112 (1971).
27. See, for example: Sienko, M.J. and Plane, R.A., *Physical inorganic chemistry*, Benjamin, New York (1963).
28. The quasi-lattice model is not mentioned in contemporary texts, though reference to it will be found in older books; in particular H.S. Frank has written on the subject.
29. Latimer, W.M., *Oxidation potentials*, Prentice–Hall, New York (1952).
30. Ree, F.H. and Hoover, W.G., *J. chem. Phys.* **40**, 939 (1964).
31. The method of obtaining Padé approximants from truncated series is illustrated in ref. 30.
32. This is the solution of the Percus-Yevick equation. Wertheim, M.S., *Phys. Rev. Lett.* **10**, 321 (1963); Thiele, E., *J. chem. Phys.* **38**, 1959 (1963).
33. Alder, B.J. and Wainwright, T.E., *Phys. Rev.* **127**, 359 (1962); also see ref. 30.
34. Guggenheim, E.A. and McGlashan, M.L., *Proc. R. Soc.* **225**, 450 (1960).
35. Landolt–Börnstein tables. 4. Teil, II. Band, p. 399. Springer–Verlag, Berlin (1961).
36. ibid., p. 412.
37. Nemethy, G. and Scheraga, H.A., *J. chem. Phys.* **36**, 3382 (1962).
38. Rahman, A. and Stillinger, F.H., *J. chem. Phys.* **55**, 3336 (1971); Stillinger, F.H. and Rahman, A., *J. chem. Phys.* **57**, 1281 (1972).

3
ASSOCIATION OF BIOPOLYMERS

Proteins and nucleic acids in water represent systems of such complexity that it is not possible to describe the conformational equilibrium of these systems starting only with the basic constants of physics. A great deal can be said, however, about the multiple equilibria for these complex systems using models containing empirical parameters. For a model to be useful and reliable it must be tested and the empirical parameters determined for molecules whose purity, molecular weight, etc. are known with great certainty. Thus before we address ourselves directly to models for naturally occurring polypeptides and polynucleotides, we turn to the study of simpler synthetic analogues. In particular we will present an analysis of the multiple equilibria in short synthetic homopolynucleotides and homopolypeptides that associate to form respectively a double-helix structure (like DNA) and a triple-helix structure (like collagen). Since these molecules are short, the molecular weight (or degree of polymerization) is known accurately, hence effects of polydispersity are eliminated (we will see that chain length has a dramatic influence on the nature of the equilibrium; hence it must be known accurately to obtain reliable information). Once the empirical parameters are determined from model synthetic systems and the statistical—mechanical model for treating the system tested, we can transfer both the empirical parameters and the insight we have gained about the nature of the equilibrium to a treatment of naturally occurring, more biologically interesting molecules.

3.1. Dimerization of Oligonucleotides

Many synthetic homopolynucleotides having a degree of polymerization of about 10 nucleotide units (oligonucleotides) associate to form a double-strand complex (S_1 representing a single-strand, S_2 a double-strand complex)

$$S_1 + S_1 \rightleftharpoons S_2. \tag{3.1}$$

One of the most important steps in constructing a model to describe such systems is to experimentally verify that eqn (3.1) indeed describes the true stoichiometry. It is not hard to imagine that short chains might form any number of complexes, perhaps even an amorphous entanglement. If theory were all-powerful, it would predict that eqn (3.1) is the dominant reaction. While I will argue in the last chapter of this book that theory can indeed provide insight about naturally occurring specific-sequence molecules that experiment cannot, at this stage our modelling theory is rather impotent and we must be certain experimentally that eqn (3.1) does describe what is going on. Since this is not a book on experimental technique we will not go into details of how to test eqn (3.1), other than to mention that the simplest tests are the measurement of molecular weight (the reaction of eqn (3.1) causes the molecular weight to double) and the analysis of mixing curves.

3.1.1. Basic equilibrium

Treating eqn (3.1) like a simple chemical equilibrium we have

$$\frac{[S_2]}{[S_1]^2} = K, \tag{3.2}$$

with the conservation of single strands,

$$[S_1] + 2[S_2] = c_0, \tag{3.3}$$

where c_0 is the total concentration of single strands in any form. It is useful to introduce the dimensionless variable[1]

$$\zeta = \frac{[S_1]}{c_0}. \tag{3.4}$$

Solving eqn (3.2) for $[S_2]$

$$[S_2] = [S_1]^2 K \tag{3.5}$$

and dividing both sides by c_0 yields

$$[S_2]/c_0 = \zeta^2 K c_0. \tag{3.6}$$

Dividing both sides of eqn (3.3) by c_0 and using eqn (3.6) to eliminate $[S_2]/c_0$ we obtain

$$\zeta + 2Kc_0\zeta^2 = 1 \text{ (conservation of single strands)}, \qquad (3.7)$$

which is a relation for the fraction of free (uncomplexed) single strands ζ.

Since K will in general be a function of temperature, eqn (3.7) shows that ζ will be both a function of temperature and total single-strand concentration c_0. If we could measure ζ, the fraction of free single strands, at a given T and c_0, then from eqn (3.7) we could calculate K. However, only in very special cases where we can use a semipermeable membrane — permeable to the free single strand and not permeable to the double strand — can we directly measure ζ; and in any case it is a difficult experiment. What we can measure easily is an optical property (circular dichroism, optical rotation, hypochromicity) that is directly proportional to the amount of ordered structure (double helix) in the complex. Once again, theory cannot predict the type of ordered structure, such as double helix, in the complex, and this must be experimentally verified (e.g. by comparing the optical properties with known examples of double helix or by X-ray studies on the solid state). If $P_0(c_0, T)$ is the optical property being studied, then we can define the fraction of ordered structure (double helix) by the relation

$$\begin{pmatrix} \text{fraction of} \\ \text{double helix} \end{pmatrix} = \vartheta(c_0, T) = \frac{|P_0(c_0, T)|}{|P_0|_{\max}}, \qquad (3.8)$$

where $|P_0|_{\max}$ is the value of the optical property when c_0 and T are such that the system has the maximum amount of ordered structure (100 per cent double helix).

3.1.2. Concentration dependence of the equilibrium[2]

I now want to show how the measurement of $\vartheta(c_0, T)$ as a function of c_0 at constant T yields K with no assumptions as to the form of K. First we note that the experimental quantity ϑ is the product of two factors

$$\vartheta = \begin{pmatrix} \text{net fraction} \\ \text{of helix} \end{pmatrix} = \begin{pmatrix} \text{fraction of} \\ \text{molecules in} \\ \text{complex} \end{pmatrix} = \begin{pmatrix} \text{fraction of} \\ \text{helix in} \\ \text{complex} \end{pmatrix}. \quad (3.9)$$

The maximum concentration of the complex (S_2) is given by

$$(\text{maximum concentration of complex}) = \tfrac{1}{2} c_0. \quad (3.10)$$

Thus the fraction of molecules in the complex is

$$(\text{fraction of molecules in complex}) = \frac{2[S_2]}{c_0}. \quad (3.11)$$

From eqn (3.6), eqn (3.11) can be given in terms of ζ as

$$(\text{fraction of molecules in complex}) = 2Kc_0 \zeta^2. \quad (3.12)$$

Without making any assumptions about the structure of K, at this stage we simply note that the fraction of helix in the complex can be obtained by appropriate (at this stage unspecified) differentiation of K which we denote as K':

$$(\text{fraction of helix in complex}) = K'. \quad (3.13)$$

Using eqns (3.13) and (3.12), eqn (3.9) becomes

$$\vartheta = 2KK'c_0 \zeta^2. \quad (3.14)$$

The experimental quantity we are investigating is the variation of ϑ with c_0 at constant temperature. It proves more useful to treat the logarithmic derivative

$$\left(\frac{\partial \ln \vartheta}{\partial \ln c_0} \right)_T = \frac{c_0}{\vartheta} \left(\frac{\partial \vartheta}{\partial c_0} \right)_T. \quad (3.15)$$

Eqn (3.14) for ϑ contains an explicit dependence on c_0 but it also contains ζ which through eqn (3.7) is a function of c_0. Thus the c_0 dependence of ϑ is given by

$$\vartheta(c_0) = 2KK'c_0 \zeta(c_0)^2, \quad (3.16a)$$

$$\zeta(c_0) + 2Kc_0 \zeta(c_0)^2 = 1. \quad (3.16b)$$

ASSOCIATION OF BIOPOLYMERS

For the example at hand, eqn (3.16b) could be solved to give ζ explicitly as a function of c_0, and this result could be substituted in eqn (3.16a) giving an explicit relation for ϑ as a function of c_0. We will encounter analogues of eqn (3.16) in Chapter 5, where eqn (3.16b) cannot be solved explicitly for ζ. Hence we present a general technique to obtain the derivative of eqn (3.16) explicitly without having to solve eqn (3.16b) explicitly; this technique incidentally is also simpler than the elimination of ζ via an explicit solution of eqn (3.16b). Recognizing the explicit and implicit dependence of ϑ and c_0 in eqn (3.16a), we can write the derivative in eqn (3.15) as

$$\left(\frac{\partial \ln \vartheta}{\partial \ln c_0}\right)_T = \left(\frac{\partial \ln \vartheta}{\partial \ln \zeta}\right)^*\left(\frac{\partial \ln \zeta}{\partial \ln c_0}\right)_T + \left(\frac{\partial \ln \vartheta}{\partial \ln c_0}\right)^*_T, \qquad (3.17)$$

where the starred derivatives in eqn (3.17) represent the explicit differentiation of eqn (3.16a) with respect to the appropriate variables.

Successively evaluating the derivatives required in eqn (3.17) we have

$$\frac{\partial \vartheta^*}{\partial \zeta} = 4KK'c_0\zeta \qquad (3.18)$$

or

$$\frac{\partial \ln \vartheta^*}{\partial \ln \zeta} = \frac{\zeta}{\vartheta}\frac{\partial \vartheta^*}{\partial \zeta} = 2. \qquad (3.19)$$

Likewise we obtain

$$\frac{\partial \ln \vartheta^*}{\partial \ln c_0} = 1. \qquad (3.20)$$

To obtain $\partial \ln \zeta/\partial \ln c_0$ we implicitly differentiate eqn (3.16b):

$$\frac{\partial \zeta}{\partial c_0} + 4Kc_0\zeta\frac{\partial \zeta}{\partial c_0} + 2K\zeta^2 = 0, \qquad (3.21)$$

which we solve for $\partial \zeta/\partial c_0$:

$$\frac{\partial \zeta}{\partial c_0} = \frac{-2K\zeta^2}{1 + 4Kc_0\zeta} \cdot \qquad (3.22)$$

Converting to the logarithmic derivatives gives

$$\frac{\partial \ln \zeta}{\partial \ln c_0} = \frac{c_0 \partial \zeta}{\zeta \partial c_0} = \frac{-2Kc_0\zeta^2}{\zeta + 4Kc_0\zeta^2}. \qquad (3.23)$$

From eqn (3.16b) we have

$$2Kc_0\zeta^2 = 1 - \zeta. \qquad (3.24)$$

Substituting eqn (3.24) in eqn (3.23) yields

$$\frac{\partial \ln \zeta}{\partial \ln c_0} = \frac{\zeta - 1}{2 - \zeta}. \qquad (3.25)$$

Inserting eqns (3.25), (3.20), and (3.19) in eqn (3.17) gives the final result

$$\left(\frac{\partial \ln \vartheta}{\partial \ln c_0}\right)_T = \frac{\zeta}{2 - \zeta}. \qquad (3.26)$$

Thus from an experimental measurement of the rate of change of ϑ as a function of c_0 at constant temperature, we can calculate ζ, the fraction of free strands. Knowing ζ, the equilibrium constant K is given by eqn (3.16b)

$$K = \frac{1 - \zeta}{2c_0\zeta^2}. \qquad (3.27)$$

By measuring the variation of ϑ with c_0 at various temperatures we obtain the thermodynamic parameters associated with K,

$$\frac{\partial \ln K}{\partial (1/T)} = -\frac{\Delta h^0}{R},$$

$$\Delta g^0 = -RT\ln K, \qquad (3.28)$$

$$\Delta s^0 = \frac{\Delta h^0 - \Delta g^0}{T}.$$

From eqn (3.28) we can study how the thermodynamic parameters depend on chain length, whether or not Δh^0 and Δs^0 are independent of temperature, etc. And we can do this without making any assumptions about the energetics of the complex or the nature of K. Note that the only assumptions we have made are eqn (3.1) (the fact that we indeed have a bimolecular complex), and eqn (3.8) (that we have an experimental measure of the amount of ordered structure in the complex).

3.1.3. Temperature dependence of the equilibrium

While the above approach is very powerful in that a bare minimum of assumptions need be made, systematic concentration studies are not common. We will give an example in Chapter 5 of the binding of oligomers to polynucleotides that utilizes the above approach. Now we pursue the more common experimental approach of studying ϑ as a function of temperature at constant c_0. The temperature variation of ϑ is most usefully expressed as

$$\frac{\partial \ln \vartheta}{\partial (1/T)} = -\frac{T^2}{\vartheta}\frac{\partial \vartheta}{T}. \tag{3.29}$$

As with eqn (3.17) the derivative of eqn (3.29) is evaluated using eqn (3.16a) where the explicit temperature dependence is through $K(T)$ and the implicit temperature dependence is through ζ which through eqn (3.16b) depends on $K(T)$. Thus

$$\frac{\partial \ln \vartheta}{\partial (1/T)} = \frac{\partial \ln \vartheta^*}{\partial \ln K}\frac{\partial \ln K}{\partial (1/T)} + \frac{\partial \ln \vartheta^*}{\partial \ln K'}\frac{\partial \ln K'}{\partial (1/T)} +$$

$$+ \frac{\partial \ln \vartheta^*}{\partial \ln \zeta}\frac{\partial \ln \zeta}{\partial \ln K}\frac{\partial \ln K}{\partial (1/T)}, \tag{3.30}$$

where once again the starred derivatives refer to the explicit differentiation of eqn (3.16a) with respect to the appropriate variables.

Proceeding as before we have

$$\frac{\partial \ln \vartheta^*}{\partial \ln K} = 1, \quad \frac{\partial \ln \vartheta^*}{\partial \ln \zeta} = 2,$$

$$\frac{\partial \ln \vartheta^*}{\partial \ln K'} = 1, \quad \frac{\partial \ln \zeta}{\partial \ln K} = \frac{\zeta - 1}{2 - \zeta}, \tag{3.31}$$

Taking the temperature coefficients of K and K' as

$$\frac{\partial \ln K}{\partial (1/T)} = -\frac{\Delta h^0}{R}, \quad \frac{\partial \ln K'}{\partial (1/T)} = -\frac{\Delta h'}{R}, \qquad (3.32)$$

then eqn (3.30) becomes

$$\frac{\partial \ln \vartheta}{\partial (1/T)} = -\frac{\Delta h^0}{R}\left\{1 + \frac{2(\zeta-1)}{2-\zeta}\right\} - \frac{\Delta h'}{R}. \qquad (3.33)$$

We have the two experimental numbers

$$\begin{aligned} A_1 &= \vartheta, \\ A_2 &= \frac{\partial \ln \vartheta}{\partial (1/T)}. \end{aligned} \qquad (3.34)$$

With eqns (3.33) and (3.16), indicating the experimental numbers with eqn (3.34), we have three relations

$$\begin{aligned} A_1 &= 2KK'c_0\zeta^2, \\ A_2 &= -\frac{\Delta h^0}{R}\left(\frac{\zeta}{2-\zeta}\right) - \frac{\Delta h'}{R}, \\ \zeta + 2Kc_0\zeta^2 &= 1, \end{aligned} \qquad (3.35)$$

and five unknowns, ζ, K, K', Δh^0, and $\Delta h'$. In general, the study of the temperature dependence of ϑ involves more parameters than the study of the concentration dependence of ϑ (with eqn (3.26) there is one equation in one unknown). Since we have more unknowns than equations, to proceed we must make some assumptions about the nature of K.

3.1.3. *Partition function for the double-strand complex*[3]

A simple model for K is obtained by considering the complex to form in successive steps as shown in Fig. 3.1. The first step represents the initial coming together of the two single strands with equilibrium constant βs. Since this step represents the loss of the translational entropy of one strand (two independent molecules are replaced by one), we expect β to be less than unity (see section 2.3.9). The equilibrium constant for adding each successive unit to the double helix is assumed to be described by the same equilibrium constant s. A given length of double helix can be located in many different ways on a single strand. Referring to Fig. 3.2, where N is the chain length (number of nucleotide units in a single strand) and n is the number of units in the double-helix conformation), we make

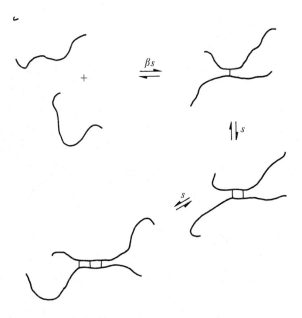

FIG. 3.1. Illustration of the parameters required to describe double-strand association of oligonucleotides.

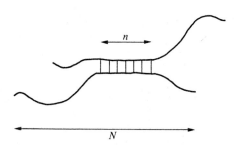

FIG. 3.2. Illustration of indexing for a double-strand complex. N is the chain length and n is the length of the sequence of double helix.

the following observations (for synthetic polynucleotides there is no reason why the two chains should be in register, i.e. any n consecutive bases in one chain can pair with any n consecutive bases in the other)

$$\begin{pmatrix}\text{number of ways } n \text{ consecutive units} \\ \text{can be picked on one strand}\end{pmatrix} = (N-n+1), \quad (3.36a)$$

$$\begin{pmatrix}\text{number of ways } n \text{ consecutive units can} \\ \text{be picked on two strands simultaneously}\end{pmatrix} = (N-n+1)^2. \quad (3.36b)$$

From Fig. 3.1, a stretch of helix n units long in a fixed position will have equilibrium constant of formation βs^n. Since from eqn (3.36b) there are $(N-n+1)^2$ different ways a stretch of helix n units long can be placed on the molecule, the total contribution of such a stretch of helix to K will be $\beta s^n (N-n+1)^2$. Finally K (a partition function that is a sum over free energy levels) is the sum over all values of n

$$K = \beta \sum_{n=1}^{N} (N-n+1)^2 s^n. \quad (3.37)$$

The probability that the double-strand complex will have n units in the double-helix conformation is given by

$$p_n = \frac{(N-n+1)^2 s^n}{\sum_{n=1}^{N} (N-n+1)^2 s^n}. \quad (3.38)$$

Note that eqn (3.38) is an example of eqn (1.77). The average value of n is then given by

$$\langle n \rangle = \sum_{n=1}^{N} n p_n \quad (3.39)$$

or

$$\langle n \rangle = \frac{\sum_{n=1}^{N} n(N-n+1)^2 s^n}{\sum_{n=1}^{N} (N-n+1)^2 s^n}. \quad (3.40)$$

ASSOCIATION OF BIOPOLYMERS

We observe that the numerator in eqn (3.40) is generated by

$$s \frac{\partial K}{\partial s} = \beta \sum_{n=1}^{N} n(N-n+1)^2 s^n, \qquad (3.41)$$

giving

$$\langle n \rangle = \frac{s}{K} \frac{\partial K}{\partial s} = \frac{\partial \ln K}{\partial \ln s}. \qquad (3.42)$$

Eqn (3.42) is to be compared with eqn (1.64) (in general if there is a one-to-one correspondence between a state (helical unit) and a factor in the partition function (s factor), then the average number of the particular states is given by the logarithmic derivative of the partition function with respect to the appropriate parameter). Writing eqn (3.9) as

$$\vartheta = \vartheta_{net} = \vartheta_{ext} \vartheta_{int}, \qquad (3.43)$$

then

$$\vartheta_{int} = \frac{\langle n \rangle}{N} = \frac{1}{N} \frac{\ln K}{\ln s}, \qquad (3.44)$$

which is an explicit relation for K' of eqn (3.13).

Eqn (3.16a) with eqn (3.44) becomes

$$\vartheta = N^{-1} 2 \zeta^2 c_0 \frac{\partial K}{\partial \ln s}. \qquad (3.45)$$

Defining the parameters (the symbol s is a standard symbol for the equilibrium constant illustrated in Fig. 3.1 and is not to be confused with the entropy per mole, Δs)

$$\beta = \exp(-\Delta h^*/RT) \exp(\Delta s^*/R), \quad \frac{\partial \ln \beta}{\partial (1/T)} = -\frac{\Delta h^*}{R},$$

$$\qquad (3.46)$$

$$s = \exp(-\Delta h_u/RT) \exp(\Delta s_u/R), \quad \frac{\partial \ln s}{\partial (1/T)} = -\frac{\Delta h_u}{R},$$

and the symbols

$$\Sigma(s) = \sum_{n=1}^{N} (N-n+1)^2 s^n ,$$

$$\Sigma(s)' = \sum_{n=1}^{N} n(N-n+1)^2 s^n , \qquad (3.47)$$

$$\Sigma(s)'' = \sum_{n=1}^{N} n^2 (N-n+1)^2 s^n ,$$

the basic equations, eqns (3.16b) and (3.45), become

$$\vartheta = 2\zeta^2 c_0 \beta \Sigma'/N , \qquad (3.48\text{a})$$

$$\zeta + 2\zeta^2 c_0 \beta \Sigma = 1. \qquad (3.48\text{b})$$

Proceeding as with eqn (3.31) we obtain

$$\frac{\partial \ln \vartheta}{\partial (1/T)} = \frac{\partial \ln \vartheta^*}{\partial \ln \zeta} \frac{\partial \ln \zeta}{\partial (1/T)} + \frac{\partial \ln \vartheta^*}{\partial \ln \beta} \frac{\partial \ln \beta}{\partial (1/T)} + \frac{\partial \ln \vartheta^*}{\partial \ln s} \frac{\partial \ln s}{\partial (1/T)}, \qquad (3.49)$$

the starred derivatives referring to explicit differentiation of eqn (3.48a). The starred derivatives are easily obtained as

$$\frac{\partial \ln \vartheta^*}{\partial \ln \zeta} = 2, \quad \frac{\partial \ln \vartheta^*}{\partial \ln s} = \frac{\Sigma''}{\Sigma'}, \quad \frac{\partial \ln \vartheta}{\partial \ln \beta} = 1. \qquad (3.50)$$

The remaining derivative is obtained by implicitly differentiating eqn (3.48b),

$$\frac{\partial \zeta}{\partial (1/T)} + 4\zeta c_0 \beta \Sigma \frac{\partial \zeta}{\partial (1/T)} + 2\zeta^2 c_0 \beta \Sigma \frac{\partial \ln \beta}{\partial (1/T)} +$$
$$+ 2\zeta^2 c_0 \beta \Sigma' \frac{\partial \ln s}{\partial (1/T)} = 0, \qquad (3.51)$$

or (using eqn (3.46) for the temperature derivatives of s and β)

$$\frac{\partial \ln \zeta}{\partial (1/T)} = \frac{\frac{\Delta h^*}{R}(2\zeta^2 c_0 \beta \Sigma) + \frac{\Delta h_u}{R}(2\zeta^2 c_0 \beta \Sigma')}{(\zeta + 2\zeta^2 c_0 \beta \Sigma) + (2\zeta^2 c_0 \beta \Sigma)} . \qquad (3.52)$$

Using eqn (3.48a) to eliminate the quantities

$$2\zeta^2 c_0 \beta \Sigma' = N\vartheta,$$

$$2\zeta^2 c_0 \beta \Sigma = N\vartheta \Sigma/\Sigma',$$

(3.53)

and recognizing eqn (3.48b) in the denominator of eqn (3.52), the final result is

$$\frac{\partial \ln \vartheta}{\partial (1/T)} = -\frac{\Delta h^*}{R}\left(1 - \frac{2}{\Sigma'/N\vartheta\Sigma + 1}\right) - \frac{N\Delta h_u}{R}\left(\frac{\Sigma''}{N\Sigma'} - \frac{2\vartheta}{1 + N\vartheta\Sigma/\Sigma'}\right).$$

(3.54)

At $\vartheta = \frac{1}{2}$

$$\vartheta' = \left(\frac{\partial \ln \vartheta}{\partial (1/T)}\right)_{\vartheta=\frac{1}{2}} = -\frac{\Delta h^*}{R}\left(1 - \frac{1}{\Sigma'/N\Sigma + \frac{1}{2}}\right) - \frac{N\Delta h_u}{R}\left(\frac{\Sigma''}{N\Sigma'} - \frac{1}{1 + N\Sigma/2\Sigma'}\right).$$

(3.55)

Note that eqn (3.55) contains three parameters: Δh^*, Δh_u, and Δs_u (the quantities in eqn (3.47) depend on s which through eqn (3.46) depends on Δh_u and Δs_u). Since eqn (3.55) contains the explicit dependence of ϑ' on chain length (note that ϑ' is roughly proportional to N), if we further assume that Δh_u, Δs_u, and Δh^* are independent of temperature (a good approximation over a limited temperature range) then the measurement of ϑ' for three different chain lengths yields the three parameters Δh_u, Δs_u, and Δh^*. The dependence of ϑ on these parameters is not simple and we must resort to a numerical search for the 'best-fit' set of parameters. We note that the sums in eqn (3.47) can be closed using the finite geometric series

$$\sum_{n=1}^{N} x^n = \frac{x^{N+1} - x}{x - 1},$$

(3.56)

$$x\frac{d}{dx}\sum_{n=1}^{N} x^n = \sum_{n=1}^{N} nx^n = \frac{x}{(1-x)^2}(Nx^{N+1} - (N+1)x^N + 1).$$

Even using eqn (3.56), the final dependence of eqn (3.55) on s is rather messy. Once the parameters Δh_u, Δs_u, and Δh^* are determined by matching the values of ϑ' for three different chain lengths, β and ζ are then given using eqn (3.48)

$$\zeta = 1 - \vartheta N \Sigma / \Sigma',$$
$$\beta = N\vartheta / 2\zeta^2 c_0 \Sigma'. \tag{3.57}$$

It should be clear that, for association reactions, the study of ϑ as a function of concentration at constant temperature greatly simplifies analysis: from eqn (3.26) we obtain ζ; then from eqn (3.51) s and β are successively determined (one parameter at a time). Repeating the experiment at different temperatures yields the parameters Δh_u, Δs_u, and Δh^* through eqn (3.46) from measurements at only two temperatures.

A useful limiting case for understanding the nature of the association of short chains is the case where the double-strand form is allowed to exist only with the maximum amount of double helix possible, as is illustrated in Fig. 3.3 (all-or-none association). For this case we have

$$\Sigma = s^N, \quad \Sigma' = Ns^N, \quad \Sigma'' = N^2 s^N,$$

$$K = \beta s^N, \quad \vartheta = 1 - \zeta, \quad \frac{\partial \ln \vartheta}{\partial \ln c_0} + \frac{1-\vartheta}{1+\vartheta}, \tag{3.58}$$

$$\frac{\partial \ln \vartheta}{\partial (1/T)} = - \left[\frac{\Delta h^*}{R} + \frac{N \Delta h_u}{R} \right] \left(\frac{1-\vartheta}{1+\vartheta} \right).$$

Evaluated at $\vartheta = \frac{1}{2}$, eqn (3.58) becomes

$$\vartheta = \zeta = \tfrac{1}{2}, \quad K = c_0^{-1},$$

$$\frac{\partial \ln \vartheta}{\partial (1/T)} = - \left[\frac{\Delta h^*}{R} + \frac{N \Delta h_u}{R} \right], \tag{3.59}$$

$$\frac{\partial \ln \vartheta}{\partial \ln c_0} = \tfrac{1}{3}.$$

ASSOCIATION OF BIOPOLYMERS 195

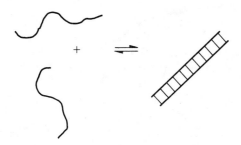

FIG. 3.3. Illustration of the all-or-none mode of chain association.

The last of eqns (3.59) provides a useful test for all-or-none behaviour.

3.1.5. *Example of oligo(A)*

For the association of oligomers of adenylic acid (N in the range 8—11) to form double helix at pH = 4, Applequist and Damle[4] obtained the following parameters:

$$\Delta h_u = -33.36 \text{ kJ mol}^{-1},$$
$$\Delta s_u = -87.36 \text{ J K}^{-1} \text{ mol}^{-1}, \quad (3.60)$$
$$\beta = 2.2 \times 10^{-3} \text{ l mol}^{-1},$$
$$\Delta h \simeq 0.$$

Fig. 3.4 shows ϑ as a function of temperature for N = 5, 10, and 15 calculated using the parameters of eqn (3.60) in eqn (3.48) with the concentration such that the total concentration of nucleotide units is constant

$$c_0(N) = c_0/N, \quad (3.61)$$

where c_0 is taken as 2.3×10^{-5} mol l^{-1}. We see that as N increases the slope of ϑ as a function of temperature becomes sharper. In addition as N is increased the temperature at which $\vartheta = \frac{1}{2}$ increases. Table 3.1 shows the values of s and ζ at $\vartheta = \frac{1}{2}$ for the different values of N used. The parameters listed in Table 3.1 yield a good deal of insight into the nature of the association equilibrium for the present system. If the

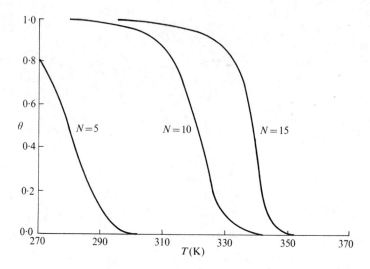

FIG. 3.4. Calculated curves for the fraction of double helix in oligo(A) for N = 5, 10, and 15; see text for parameters used.

TABLE 3.1

Parameters for oligo(A) association at ϑ = ½

N	s	ζ
5	39.0	0.49
10	6.2	0.47
15	3.5	0.45

equilibrium were all-or-none, as illustrated in Fig. 3.3, then the only mechanism for loss of helical content is the dissociation of the double-strand complex (as opposed to unravelling from the ends which is ruled out in the all-or-none model). In this case, ϑ is a direct measure of ζ (eqn (3.58)) and at $\vartheta = \frac{1}{2}$, $\vartheta = \zeta = \frac{1}{2}$. From Table 3.1 (the calculations allowing all stages of unwinding) we see that indeed ζ is very close to $\frac{1}{2}$. Thus the dominant mechanism for the loss of helical content in this system (for the range of N shown) is by simple chain dissociation as shown in Fig. 3.3 and not by unwinding from the ends. To understand why this is so we examine the values of s at $\vartheta = \frac{1}{2}$ which are seen to be significantly larger than unity. From Fig. 3.1, s is the equilibrium constant for adding a helical unit to an already existing run of helix. If s is significantly greater than unity then the all-helix state is favoured. Table 3.2 shows p_n of eqn (3.38) for $N = 10$ using the appropriate value of s for $N = 10$ and $\vartheta = \frac{1}{2}$ from Table 3.1. We see that the all-helix state, $n = N$, is strongly favoured, there being increasingly small probabilities for $n < N$. To understand why s is greater than unity when $\vartheta = \frac{1}{2}$, note from eqn (3.59) that when $\vartheta = \frac{1}{2}$,

$$c_0 \beta s^N = 1. \tag{3.62}$$

Since the quantity $c_0\beta$ has the value 5×10^{-8}, s must be significantly larger than unity to obtain the condition of eqn (3.62) for ϑ to be $\frac{1}{2}$. In turn the reason why $c_0\beta$ is significantly less than unity is that this combination of factors reflects the loss of translational entropy on the association of two independent single strands to form the double strand complex. Since β is found experimentally not to vary noticeably

TABLE 3.2.

Probability of a helical sequence n units long in the poly(A) double-strand helix with N=10 (the concentration is such that $\vartheta=\frac{1}{2}$ at each temperature)

n	p_n (300 K)	p_n (320 K)	p_n (340 K)
1	0.000	0.000	0.001
2	0.000	0.000	0.001
3	0.000	0.000	0.004
4	0.000	0.000	0.010
5	0.000	0.001	0.024
6	0.000	0.006	0.055
7	0.003	0.028	0.117
8	0.030	0.106	0.215
9	0.201	0.319	0.314
10	0.766	0.539	0.258

with chain length, it is clear that as N increases, the value of s at $\vartheta = \frac{1}{2}$ decreases toward unity,

$$s(\vartheta=\tfrac{1}{2}) \sim \left(\frac{1}{\sigma_0 \beta}\right)^{1/N} \sim 10^{8.3/N}. \tag{3.63}$$

A simple diagnostic test for all-or-none-like behaviour is to examine the concentration dependence of ϑ as a function of temperature. If the association is approximately all-or-none, then $\vartheta(T)$ will simply be shifted to lower temperature as the concentration is decreased without a noticeable change in the shape of the curve. This is illustrated in Fig. 3.5, where $\vartheta(T)$ is calculated for $N = 10$ at several concentrations. As we have already emphasized, the experimental realization of the curves shown in Fig. 3.5 is sufficient to determine all the parameters of the model for $N = 10$ independent of other chain lengths. The analysis of experimental curves similar to those shown in Fig. 3.4 requires a tedious numerical search for the set of parameters that satisfy eqn (3.55) for three different chain lengths.

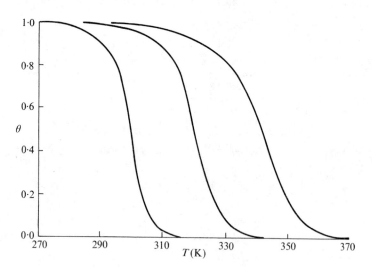

FIG. 3.5. Calculated curves for the fraction of double helix in oligo(A) with $N = 10$ at different concentrations. As the system is made more dilute the curves are simply shifted to lower temperatures.

ASSOCIATION OF BIOPOLYMERS

In the limit as N becomes large the temperature where $\vartheta = \frac{1}{2}$ increases and the curves become sharper. As N becomes very large the melting curves become infinitely steep but approach a limiting melting temperature

$$\begin{aligned}\vartheta &= 1 \ (s < 1), \\ \vartheta &= 0 \ (s > 1).\end{aligned} \qquad (3.64)$$

Assuming Δh_u and Δs_u are independent of temperature, then the temperature T_0 where $s = 1$ is given by (at $s = 1$, $\Delta g_u = \Delta h_u - \Delta s_u = 0$),

$$T_0 = \frac{\Delta h_u}{\Delta s_u} \simeq 382 \text{ K}. \qquad (3.65)$$

Using eqn (3.65) to eliminate Δs_u in favour of T_0, eqn (3.46) for s becomes

$$s = \exp\{\Delta h_u (T-T_0)/RTT\}. \qquad (3.66)$$

A useful result for all-or-none-like behaviour is obtained from the general mathematical relation

$$\left(\frac{\partial \ln \vartheta}{\partial (1/T)}\right)_{c_0} \left(\frac{\partial (1/T)}{\partial \ln c_0}\right)_{\vartheta} \left(\frac{\partial \ln c_0}{\partial \ln \vartheta}\right)_T = -1 \qquad (3.67)$$

yielding

$$\left(\frac{\partial \ln c_0}{\partial (1/T)}\right)_{\vartheta} = -\left(\frac{\partial \ln \vartheta}{\partial (1/T)}\right)_{c_0} \bigg/ \left(\frac{\partial \ln \vartheta}{\partial \ln c_0}\right)_T. \qquad (3.68)$$

For all-or-none behaviour at $\vartheta = \frac{1}{2}$, eqn (3.59) using eqn (3.68) gives

$$\left(\frac{\partial \ln c_0}{\partial (1/T)}\right)_{\frac{1}{2}} = \left(\frac{\Delta h^*}{R} + \frac{N\Delta h_u}{R}\right). \qquad (3.69)$$

Assuming Δh^* and Δh_u to be temperature independent, a plot of the derivative in eqn (3.69) versus chain length yields

$$\text{slope} = \Delta h_u/R,$$
$$\text{intercept} = \Delta h^*/R, \quad (3.70)$$

assuming all-or-none behaviour.

Since our purpose is to obtain parameters for use in naturally occurring macromolecules, we forgo an analysis of intermediate-chain-length synthetic polynucleotides and turn to another short-chain system, the analysis of which yields a great deal of information about appropriate models and parameters for naturally occurring nucleic acids.

3.2. Poly(A–T)

The synthetic polynucleotide having the regular sequence (A = adenine, T = thymine) ... ATATAT ... is generated spontaneously[5] by enzymatic synthesis *in vitro*. Degradation of the very long chains formed followed by careful separation of the fragments gives short chains of well-defined chain length. With this system it has been possible to study most of the features present in a realistic model for nucleic acids.[6] We start our discussion of this system by considering the dimerization of short chains, applying the technique of the last section.

3.2.1. Dimerization of short chains

Short chains of oligo(A–T) form double-strand complexes as illustrated in Figs. 3.1 and 3.2. For this system the double helix is of the Watson–Crick variety characteristic of DNA, unlike the double helix in poly(A) which generates a helix of much smaller radius than in DNA, being unique to poly(A). The treatment of oligo(A–T) association is exactly analogous to that given in section 3.1, with one minor difference. A pairs with T but A does not pair with A (in the Watson–Crick scheme) nor T with T. Thus if we pick a consecutive run of n units in one chain, there are only $\frac{1}{2}(N-n+1)$ ways to pick a sequence of n units in the other chain that will match A with T. Thus eqn (3.37) is given by

$$K = \tfrac{1}{2}\beta \sum_{n=1}^{N} (N-n+1)^2 s^n. \quad (3.71)$$

The thermodynamic parameters for poly(A–T) are

$$\Delta h_u = -31\,770 \text{ J mol}^{-1},$$
$$\Delta s_u = -92.8 \text{ J K}^{-1} \text{ mol}^{-1}. \qquad (3.72)$$

The temperature where $s = 1$ (see eqn (3.65)) is given by

$$T_0 = \Delta h_u/\Delta s_u = 342.5 \text{ K}. \qquad (3.73)$$

The melting behaviour of short chains of poly(A–T) is very similar to that for short chains of poly(A) given in section 3.1.5.

One interesting feature of poly(A–T) is that the chain can bend back on itself in hairpin fashion forming intramolecular double helix. If the temperature of a solution of dissociated short chains is quickly lowered, then intramolecular double helix forms more rapidly than intermolecular double helix. This is illustrated schematically in Fig. 3.6. Because of the small

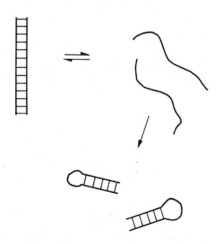

FIG. 3.6. Possible equilibria in poly(A–T). The double-strand complex is in equilibrium with the single strands; if the temperature is rapidly lowered the single strands can be trapped in the form of intra-strand double helix, giving rise to a hairpin structure.

loop required for the chain to bend back on itself, not as much double helix can form in the hairpin. Knowing the optical properties of the interchain double helix, we can estimate that it takes 4–5 units to make the turn in the hairpin (we simply estimate how much less helix the hairpin has at low temperature than the interchain form). The intrinsic stability (hairpin or two-strand complex) depends on whether the translational entropy loss (dependent on total strand concentration) due to bimolecular association in the complex or the lost section of helix in the bend in the hairpin are greater. It is found under typical experimental conditions that the bimolecular complex is more stable.

The fact that poly(A–T) can fold back on itself forming Watson–Crick double helix has led to a fascinating set of experiments[6] that has yielded a great deal of basic information about nucleic acids. It is possible to cyclize short chains of poly(A–T). In these cyclic systems double helix must form intramolecularly, as illustrated in Fig. 3.7. Since these structures are cyclic we must first consider the properties such loop molecules have before we can construct a model to analyse the behaviour of the equilibrium shown in Fig. 3.7.

FIG. 3.7. Illustration of double-helix formation in cyclic poly(A–T).

ASSOCIATION OF BIOPOLYMERS 203

3.2.2. Properties of loops

For a polymer like polyethylene there are, to a first approximation, three well-defined positions of internal rotation about each C—C bond (see Fig. 2.5, p. 124). Thus the number of conformations of a polyethylene chain of N CH_2 units is given by

$$\text{total number of conformations} = \Omega \simeq \omega^N, \quad (3.74)$$

where $\omega = 3$ for polyethylene. We will see in Chapter 4 that even if all of the positions of internal rotation are not equivalent and even if we include a vibrational partition function for each position of internal rotation, the partition function for a random chain-like polyethylene (so-called random coil: the collection of all the possible conformations of the molecule) can be written as

$$q_{\text{free chain}} = \text{constant} \times u^N, \quad (3.75)$$

where u is the average partition function per unit.

The Ω conformations of a chain can be catalogued according to how many conformations have a given distance between the ends of the molecule (end-to-end distance). The nature of the distribution of end-to-end distances can be obtained from a very simple model. Consider a linear chain of N links in one dimension where there are two orientations ($\omega = 2$) of each link, the positive and negative directions. There are then

$$\Omega = 2^N \quad (3.76)$$

conformations. If N_+ and N_- are the number of links oriented in the positive and negative directions respectively then we have

$$N = N_+ + N_-,$$
$$n = N_+ - N_-, \quad (3.77)$$

where n is the end-to-end distance. The number of ways we can arrange N_+ and N_- links is

$$\Omega(N_+, N_-) = \frac{N!}{N_+! N_-!}. \tag{3.78}$$

Using eqn (3.77) to eliminate N_+ and N_- in favour of N and n gives an expression for the number of conformations having an end-to-end distance of n links

$$\Omega_n = \frac{N!}{\left(\frac{N+n}{2}\right)! \left(\frac{N-n}{2}\right)!}. \tag{3.79}$$

Evaluating Ω_n for small N reveals that Ω_n is a maximum at $n = 0$ and quickly falls off (symetrically) for $n < 0$ and $n > 0$. Expanding ln Ω_n in a Taylor series about $n = 0$ (the first derivative being zero since Ω_n is a maximum at $n = 0$) and keeping the first non-zero term beyond the constant term, we have

$$\Omega_n \simeq \text{constant} \times \exp(-n^2/2N). \tag{3.80}$$

The probability that a chain has an end-to-end distance is then given by

$$p_n = \Omega_n/\Omega, \tag{3.81}$$

which has the same dependence on n as eqn (3.80) (simply a different constant).

In three dimensions we find

$$\Omega(n_x, n_y, n_z) = \text{constant} \times \exp(-(n_x^2 + n_y^2 + n_z^2)/\alpha N). \tag{3.82}$$

Introducing

$$r^2 = n_x^2 + n_y^2 + n_z^2$$

and summing over all values of n_x, n_y, and n_z that give the same value of r, we find (α a constant characteristic of the polymer)

$$p_r = \text{constant} \times r^2 \exp(-r^2/\alpha N). \tag{3.83}$$

We can normalize p_r by requiring

$$\int_0^\infty p_r \, dr = 1. \qquad (3.84)$$

The upper limit used in eqn (3.84) is an approximation: the largest value r can have is the maximum contour length of the chain (proportional to N), but since the maximum in p_r is proportional to $N^{\frac{1}{2}}$ and the function falls off rapidly, taking the limit to ∞ is justified. Using eqn (3.84) we obtain

$$p_r = \text{constant} \times N^{-\frac{3}{2}} \exp(-r^2/\alpha N) r^2. \qquad (3.85)$$

Returning to loops, from eqn (3.85) the probability of a chain returning to the origin ($r = \Delta$, Δ a small increment near $r = 0$) is given by

$$p_{\text{loop}} = \frac{\text{constant}}{N^{-3/2}}, \qquad (3.86)$$

where the constant is a characteristic parameter for the particular polymer. The number of conformations that a loop has is then given by

$$\Omega_{\text{loop}} = \Omega_{\text{free chain}} \times \frac{\text{constant}}{N^{-3/2}}. \qquad (3.87)$$

It is clear from eqn (3.85) that the fraction of conformations that form a loop are only a small fraction of the total number of conformations, i.e. $\Omega_{\text{loop}}/\Omega_{\text{free chain}} \ll 1$. With eqn (3.87) the analogue of eqn (3.75) for a loop is

$$q_{\text{loop}} = au^N/N^c, \qquad (3.88)$$

where a is a constant. Implicit in the derivative of eqn (3.85) was the condition that conformations which brought distant segments of the chain to the same location in space were not ruled out (excluded volume was ignored). Computer studies[7] on model polymers where the effect of excluded volume is taken into account yields

$$c = 1.5 \text{ (excluded volume not taken into account)}, \quad (3.89)$$
$$c = 1.75 \text{ (excluded volume taken into account)},$$

where the value $c = 1.75$ is found independent of the nature of the model polymer.

3.2.3. Statistical weights for nucleic acids[8]

We now want to give a very specific assignment of statistical weights for the system in Fig. 3.7. While I will argue later that this assignment is not the most general approach we can take, it is instructive to see a concrete assignment in terms of specific molecular properties and interactions. The term 'statistical weight' requires some comment. It is clear that the number of conformations of a biological macromolecule in solution is enormous. Thus the partition function (sum over appropriate free energy levels) sums over a very large number of terms. If we were to assign unknown parameters (to be determined by comparison of theory with experiment) to each important conformation, there clearly would not be enough experimental information available. As illustrated in section 3.1, we must, of necessity, write all of the required free energies in terms of a few parameters such as Δh_u, Δs_u, Δh^*, and Δs^* of section 3.1. The few component factors out of which the partition function for the total system is constructed are commonly called statistical weights. Thus in section 3.1 the partition function was constructed using the two statistical weights s and β.

For the system in Fig. 3.7 we assume that any combination of loops and double helix can be described in terms of the following statistical weights

State of a unit in the molecule	Statistical weight	
Backbone unit in a single chain (no loop)	u	
Two bases hydrogen bonded (Watson–Crick pairing)	t	(3.90)
Two neighbouring pairs of hydrogen-bonded bases (stacking interaction)	τ.	

ASSOCIATION OF BIOPOLYMERS

The statistical weight u largely reflects the conformational entropy of a unit in an unrestricted chain. The statistical weight t largely reflects the energy of hydrogen bonding between two bases (although we must remember that in solution this is the difference in free energy between the base—base hydrogen bonds when the Watson—Crick pair is formed and the base—water hydrogen bonds when it is broken — see section 2.2.12). The statistical weight τ reflects the fact that a hydrogen-bonded base pair presents a large flat surface; when two of these base pairs exist side by side in the molecule (as they do successively in the double helix) then at least a portion of the flat surfaces overlap, giving rise to a complicated and important inter-base pair interaction (consisting of electrostatic interactions between the polar bases and changes in the solvent structure — see section 2.2.13).

With the statistical weights of (3.90) plus the form of eqn (3.88), representing a correction of the statistical weight u when a chain is in a loop (reduction of conformational entropy), we can formulate the contribution of any structure such as shown in Fig. 3.7 to the partition function. For example, in Fig. 3.8 four different states of the molecule are shown. In terms of the statistical weights and the loop entropy just

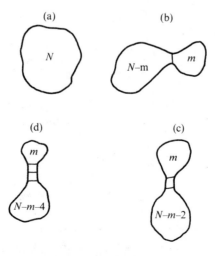

FIG. 3.8. Illustration of indexing of loops in cyclic poly(A—T).

described we have (where N is the total number of nucleotide units)

Structure in Fig. 3.8	Contribution of the partition function	
a	au^N/N^c	
b	$a^2u^Nt/\{(N-m)^c m^c\}$	(3.91)
c	$a^2u^{N-2}t\tau/\{(N-m-2)^c m^c\}$	
d	$a^2u^{N-4}t^2\tau^2/\{(N-m-4)^c m^c\}$	

The generalization of (3.91) is

$$\begin{Bmatrix}\text{Contribution to the}\\\text{partition function of a}\\\text{structure with } n \text{ base}\\\text{pairs in double helix,}\\\text{one loop containing } m\\\text{chain units the other}\\\text{containing } (N-m-2n)\end{Bmatrix} = \frac{a^2(u/\tau)^2(t\tau/u^2)^n u^N}{(N-m-2n)^c m^c}. \quad (3.92)$$

The partition function then is a sum of all such terms over all values of n and m. The partition function is

$$q = \frac{au^N}{N^c} + a^2\left(\frac{u^2}{\tau}\right)u^N \sum_m \sum_n \left(\frac{t\tau}{u^2}\right)^n / \{(N-m-2n)^c m^c\} \quad (3.93)$$

$$= \frac{au^N}{N^c}\left\{1 + a\frac{u^2}{\tau}\sum_m\sum_n\left\{\frac{N}{(N-m-2n)m}\right\}^c\left(\frac{t\tau}{u^2}\right)^n\right\}.$$

where the sums are over all allowed values of m and n (the details of which we forgo here). Defining

$$\sigma = au^2/\tau,$$
$$s = t\tau/u^2, \quad (3.94)$$

then the average number of hydrogen-bonded base pairs in the molecule is given in the standard manner (see eqn (3.42)),

$$\langle n \rangle = \frac{\partial \ln q}{\partial \ln s}$$

(3.95)

$$= \frac{\sigma \sum_m \sum_n n \left\{\frac{N}{(N-m-2n)m}\right\}^c s^n}{1 + \sigma \sum_m \sum_n \left\{\frac{N}{(N-m-2n)m}\right\}^c s^n}.$$

The important point about eqn (3.95) is that only the parameters σ and s appear and that it is impossible to obtain separately the parameters τ, t, u^2, and a. We note that σ and s can be thought of as the equilibrium constants for the processes

$$\begin{pmatrix}\text{two units in} \\ \text{a free chain}\end{pmatrix} \underset{}{\overset{s}{\rightleftharpoons}} \begin{Bmatrix}\text{hydrogen-bonded base pair} \\ \text{followed by a hydrogen-bonded} \\ \text{base pair}\end{Bmatrix}$$

(3.96)

$$(\text{free chain}) \underset{}{\overset{\sigma}{\rightleftharpoons}} \begin{Bmatrix}\text{nucleation of an isolated hydrogen-} \\ \text{bonded base pair resulting in the} \\ \text{formation of a new loop}\end{Bmatrix}.$$

Thus it is only the difference in free energy (ratio of statistical weights, s, $= t\tau/u^2$) that ever appears in an expression for an average quantity (see section 1.1). In the literature we often find σ for nucleic acids equated with τ^{-1}, thus giving an estimate of the stacking parameter. From eqn (3.94) this is quite incorrect.

In Chapter 4 we will see that eqns (3.90), (3.91), and (3.94) are overly restrictive interpretations of σ and s and will argue for the general form such as given in eqn (3.96). The parameters Δh_u and Δs_u which determine s (see eqn (3.46)) have already been given in eqn (3.72). Fitting the experimental data on the melting curves for cyclic poly(A–T) with eqn (3.95) yields the value of

$$\sigma = 0.003-0.004.$$

(3.97)

This is a very important parameter since it is the statistical weight for nucleating a new loop in the double helix. Incidentally, the factors m^{-c}, etc., in eqn (3.95) prevent a closed form from being obtained; hence we must perform the summations explicitly on the computer.

3.3. Association of $(Gly-Pro-Pro)_N$

Collagen consists of three parallel and equivalent polypeptide chains arranged in a triple helix with inter-strand hydrogen bonds between the chains. Sequence studies (see section 3.5) show that the polypeptide chains in collagen have the general structure $(Gly-x-y)_N$, i.e. every third residue is glycine followed by any two other amino or imino acids. A particular collagen is characterized by the specific sequence of xs and ys. The average amino-acid composition of collagens is approximately one-third glycine, one-third proline and hydroxyproline, and one-third of all the other amino acids. It has been found that synthetic analogues having a regular $(Gly-x-y)_N$ structure also form a triple-helix complex having the basic properties of the collagen triple helix. As with polynucleotides, the study of well-characterized synthetic analogues allows us to obtain parameters for and test models of natural collagen. One such system that has been carefully studied is $(Gly-Pro-Pro)_N$.

3.3.1. Triple-helix equilibrium[9]

Fig. 3.9 illustrates the association of three strands of $(Gly-Pro-Pro)_N$ to form triple helix with the assignment (in analogy with Fig. 3.1) of equilibrium constants βs (for the initial joining of the strands) and s (for the addition of a tripeptide unit from each chain to the triple helix). The treatment of this system is exactly analogous to that given in section 3.1, except that three strands are involved. Thus we have the equilibrium

$$3S_1 \rightleftharpoons S_3, \qquad (3.98)$$

with the constraint

$$[S_1] + 3[S_3] = c_0, \qquad (3.99)$$

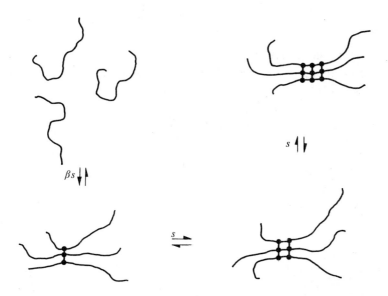

FIG. 3.9. Illustration of the parameters required to describe triple-strand association in $(Gly-Pro-Pro)_N$.

c_0 again being the total concentration of single strands in any form. The analogue of eqn (3.7) (following the same development) is

$$\zeta = 3Kc_0^2\zeta^3 = 1, \qquad (3.100)$$

where ζ is defined in eqn (3.4) (the fraction of free single strands). The analogue of eqn (3.26) (again derived in the same fashion) is

$$\left(\frac{\partial \ln \vartheta}{\partial \ln c_0}\right)_T = \frac{2\zeta}{3 - 2\zeta}. \qquad (3.101)$$

Knowledge of ζ from eqn (3.101) (by experimental measurement of the derivative) yields K from eqn (3.100); determination of K at different temperatures yields all the thermodynamic parameters via eqn (3.28).

In terms of β and s, using Fig. 3.9, K is given by

$$K = \beta \sum_{n=1}^{N} (N-n+1)^3 s^n, \qquad (3.102)$$

where N is the chain length (number of Gly–Pro–Pro triplets) and n is the number of triplets per chain in the triple helix. With eqn (3.102), the analogue of eqn (3.48) is

$$\vartheta = 3\zeta^3 c_0^2 \beta \Sigma'/N,$$
$$\zeta = 3\zeta^3 c_0^2 \beta \Sigma = 1; \qquad (3.103)$$

$$\Sigma(s) = \sum_{n=1}^{N} (N-n+1)^3 s^n,$$
$$\Sigma(s)' = \sum_{n=1}^{N} n(N-n+1)^3 s^n, \qquad (3.104)$$
$$\Sigma(s)'' = \sum_{n=1}^{N} n^2 (N-n+1)^3 s^n.$$

Following the derivation of eqn (3.55) using eqns (3.103) and (3.104) with the definition of β and s in eqn (3.46) we have

$$\vartheta' = \left[\frac{\partial \ln \vartheta}{\partial (1/T)}\right]_{\vartheta=\frac{1}{2}} =$$

$$= -\frac{\Delta h^*}{R}\left(1 - \frac{3/2}{1 + \Sigma'/N\Sigma}\right) - \frac{N\Delta h_u}{R}\left(\frac{\Sigma}{N\Sigma'} - \frac{3/2}{1 + \Sigma'/\Sigma'}\right). \qquad (3.105)$$

The average length and average length squared of triple helix are given by

$$\langle n \rangle = \Sigma'/\Sigma,$$
$$\langle n_2 \rangle = \Sigma''/\Sigma. \qquad (3.106)$$

Using eqn (3.106), eqn (3.105) has the form

$$\vartheta' = -\frac{\Delta h^*}{R}\left(\frac{\langle n\rangle/N - \frac{1}{2}}{\langle n\rangle/N + 1}\right) - \quad (3.107)$$

$$- \langle n\rangle \frac{\Delta h_u}{R}\left(\frac{\langle n^2\rangle/N\langle n\rangle + \langle n^2\rangle/\langle n\rangle^2 - 3/2}{\langle n\rangle/N + 1}\right).$$

As with short polynucleotides, the all-or-none limit is useful (dissociated single strands in equilibrium with 100 per cent triple helix). Allowing only the triple-helix state ($n = N$) we have

$$K = \beta s^N = \vartheta/3c_0^2(1-\vartheta)^3,$$

$$\vartheta = 1 - \zeta,$$

$$\left(\frac{\partial \ln \vartheta}{\partial \ln c_0}\right)_T = 2(1-\vartheta)/(1+2\vartheta), \quad (3.108)$$

$$\left(\frac{\partial \ln \vartheta}{\partial (1/T)}\right)_{c_0} = -\left(\frac{\Delta h^*}{R} + \frac{N\Delta h_u}{R}\right)\left(\frac{1-\vartheta}{1+2\vartheta}\right).$$

At $\vartheta = \frac{1}{2}$ eqn (3.108) becomes

$$K = \frac{4}{3}c_0^2,$$

$$\left(\frac{\partial \ln \vartheta}{\partial \ln c_0}\right)_T = \frac{1}{2}, \quad (3.109)$$

$$\left(\frac{\partial \ln \vartheta}{\partial (1/T)}\right)_{c_0} = -\frac{\Delta h^*}{4R} - \frac{N 4 h_u}{4R}.$$

3.3.2. Thermodynamic parameters

The discrete points in Fig. 3.10 show the experimental data of Kobayashi and co-workers[10] for $N = 10$, 15, and 20, while the solid curves represent calculated melting curves obtained from eqn (3.103). The first parameter that is determined is T_0 which

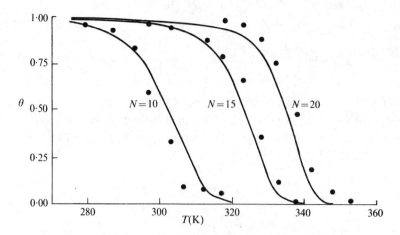

FIG. 3.10. Melting curves for $(Gly\text{–}Pro\text{–}Pro)_N$, $N = 10$, 15, and 20. The dots are experimental points while the solid curves are calculated.

is obtained by plotting $T_{\frac{1}{2}}$ (the temperature at which $\vartheta = \frac{1}{2}$) or $1/T_{\frac{1}{2}}$ as a function of $1/N$. The extrapolated value (melting temperature of infinite chains where $\vartheta = \frac{1}{2}$ at $s = 1$; see eqn (3.64)) that is obtained is

$$T_0 = 376 \text{ K}. \tag{3.110}$$

Using eqn (3.105) for ϑ', we have three experimental numbers from the data of Fig. 3.10. A 'best-fit' search gives

$$\begin{aligned}\Delta h^* &\sim 0, \\ \Delta s^* &= -78 \pm 8 \text{ J mol}^{-1} \text{ K}^{-1}, \\ \Delta h_u &= -30.9 \pm 3.3c \text{ kJ mol}^{-1}, \\ \Delta s_u &= -84 \pm 8 \text{ J mol}^{-1} \text{ K}^{-1}.\end{aligned} \tag{3.111}$$

The parameter β (found to be independent of temperature) has the value

$$\ln \beta = -9.3 \pm 2.0. \tag{3.112}$$

Note that β has the units of $(1 \text{ mol}^{-1})^2$ which are the units of a trimolecular association constant.

3.3.3. Properties of the equilibrium

As with the association of oligonucleotides to form double helix, the association of short chains of (Gly–Pro–Pro)$_N$ to form the triple helix involves the loss of translational entropy (in the present case three independent chains are replaced by one complex, hence the translational entropy of two independent molecules is lost). For the triple helix to be stable even with this loss of translational entropy, the factor s must be larger than unity. Again, as with the oligonucleotide case, $s > 1$ implies that the complex will exist approximately in the 100 per cent helical form, states with significant unwinding having a very low probability. The probability a complex will have a triple-helix sequence n units long is given by

$$p_n = \frac{(N-n+1)^3 s^n}{\Sigma (N-n+1)^3 s^n}. \tag{3.113}$$

Fig. 3.11 shows p_n for $N = 10$, 15, and 20 at various temperatures through the transition region for each chain length. We clearly see that while the distribution of helix lengths becomes broader as the chain length increases (as N increases, the temperature where $\vartheta = \frac{1}{2}$ increases and s decreases toward unity; $\beta c_0^2 s^N \simeq 1$ is the criterion for $\vartheta = \frac{1}{2}$, and as N increases the value of s required to meet this condition decreases). The probability of any significant fraction of unravelling from the ends is very small.

From Fig. 3.11, the nature of the equilibrium is very close to the all-or-none behaviour. Having determined the parameters of eqn (3.111) from a study of experimental curves for $N = 10$, 15, and 20, we can now use these parameters in eqn (3.103) and calculate the behaviour of much longer chains. The first calculation we present is the value of ζ at $\vartheta = \frac{1}{2}$. From eqn (3.108), if the equilibrium is strictly all or none, then $\vartheta = 1 - \zeta$ or $\zeta = \frac{1}{2}$ when $\vartheta = \frac{1}{2}$. The values of ζ at $\vartheta = \frac{1}{2}$ are shown in Table 3.3 for chain lengths (number of Gly–Pro–Pro units in a chain) up to $N = 300$. We see that even at $N = 300$ the

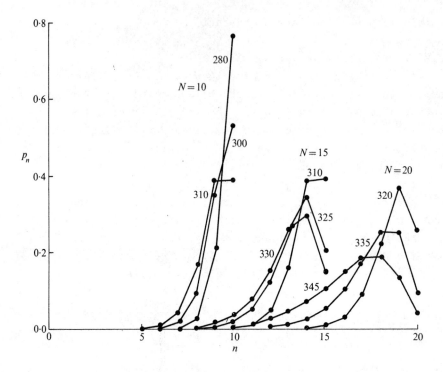

FIG. 3.11. Probability of having a triple-helix sequence of n units in (Gly–Pro–Pro)$_N$ as a function of chain length and temperature.

TABLE 3.3.

The fraction ζ of free strands as $\vartheta = \frac{1}{2}$ as a function of chain length for the triple-strand association of (Gly–Pro–Pro)$_N$

10	0.463
15	0.445
20	0.434
100	0.392
300	0.373

ASSOCIATION OF BIOPOLYMERS

mechanism by which helical content decreases is still largely through unwinding from the ends. This is further illustrated in Fig. 3.12, where the distribution p_n at $\vartheta = \frac{1}{2}$ is calculated for $N = 50$, $N = 100$, and $N = 300$. Again, as for $N = 10$, 15, and 20 (Fig. 3.11), the distribution becomes broader as N increases, but there is at most minor fraying from the ends and little probability of structured with very short helical segments.

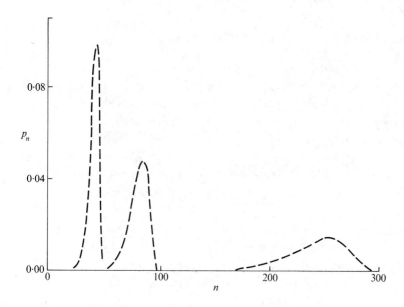

FIG. 3.12. Probability of having a triple-helix sequence of n units in $(Gly-Pro-Pro)_N$ for chain lengths $N = 50$, 100, and 300 at a temperature such that $\vartheta = \frac{1}{2}$ for the respective chain lengths.

It must be emphasized that the above calculations for large N do not introduce the possibility of unwinding internally via the formation of internal loops. Such calculations will be explored in Chapter 6. The results of these calculations show that interior unwinding plays no role in the equilibrium for the chain lengths treated here.

3.4. Statistical weights for natural collagen

In the last section we obtained thermodynamic parameters for triplex helix formed from (Gly–Pro–Pro)$_N$ chains and studied the nature of the equilibrium in this system. Since natural collagen contains many different triplets of the general structure (Gly–x–y), we inquire here as to whether we can obtain any additional information about parameters for natural collagen from the study of (Gly–Pro–Pro)$_N$. To begin we examine the dependence of the melting temperature of natural collagen on amino-acid composition.

3.4.1. Stability of collagen as a function of amino-acid composition

The average amino-acid composition of collagen (approximately one-third glycine, one-third imino acids (proline and hydroxyproline), and one-third of the other amino acids) is intriguing. From gross steric considerations we can classify the amino acids into three categories,[11] as illustrated in Fig. 3.13. The different conformations of a polypeptide chain arise from internal rotation about the dihedral angles φ and ψ indicated in Fig. 3.13 (internal rotation about the dihedral angle ω of the amide group is severely hindered since, owing to resonance stabilization and steric effects, this group prefers the *trans* planar conformation, the barrier between the two planar conformations, *cis* and *trans*, being about 8.5 kJ mol^{-1}). Clearly the imino acids proline and hydroxyproline have the least conformational freedom, the angle φ being fixed by the five-membered ring and there being only a small range of allowed values of ψ due to steric hindrance. Since glycine has no β carbon, it has the greatest conformational freedom. All the rest of the amino acids have about the same conformational freedom of the backbone, the structure of the side-chain beyond the β carbon not having a major influence on steric considerations. Thus according to steric considerations collagen has the average composition: one-third most restricted (Pro and Hypro); one-third least restricted (Gly); and one-third intermediate (all the rest of the amino acids).

Since from the general formula (Gly–x–y) every third residue is Gly, the amount of this amino acid is constant (almost — occasionally x or y can be Gly). Thus we would expect the thermal stability of collagen (as measured by its melting

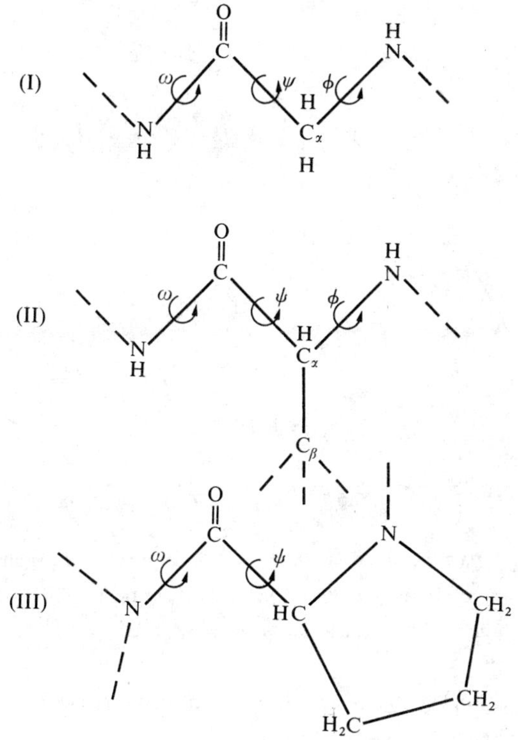

FIG. 3.13. Illustration of the types of amino-acid residues in polypeptides: (I) glycine with no β carbon; (III) proline (or hydroxyproline) with the angle φ essentially frozen by the five-membered ring; (II) all the rest of the amino acids (having a β carbon).

temperature T_m, where $\vartheta = \frac{1}{2}$) to be a function mainly of the imino-acid content (the conformation entropy of the dissociated strands being to first approximation inversely proportional to the fraction of imino acids present, thus increasing the stability of the triple helix relative to the random coil). This indeed is found to be the case. The solid dots in Fig. 3.14 show the melting temperature of various natural collagens as a function of the fraction α of imino-acid residues; the open circle at $\alpha = \frac{2}{3}$ is the melting temperature for very long (Gly–Pro–Pro)$_N$ (eqn (3.110)) (since we have seen T_m is very dependent on chain length and concentration for short chains,

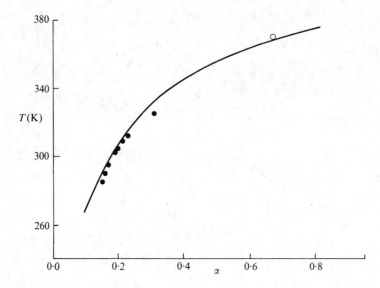

FIG. 3.14. Melting temperatures ($\vartheta = \frac{1}{2}$) of natural collagens as a function of the fraction α of imino acids. The open circle is for $(Gly-Pro-Pro)_N$, N very large. The curve is calculated (see text).

we want to use only T_m for very long chains where the T_m occurs approximately when $\langle s \rangle = 1$).

3.4.2. Simple model for the melting temperature

For a first approximate assumption we take the melting temperature of a natural collagen to be given by

$$T_m(\alpha) = \Delta h_u(\alpha)/\Delta s_u(\alpha) \tag{3.114}$$

with

$$\Delta h_u(\alpha) = 9\{(1-\alpha)\Delta h_a + \alpha\Delta h_i\},$$
$$\Delta s_u(\alpha) = 9\{(1-\alpha)\Delta s_a + \alpha\Delta s_i\}, \tag{3.115}$$

where the subscripts a and i refer to the contribution of amino and imino acids respectively. The parameters $\Delta h_u(\alpha)$ and $\Delta s_u(\alpha)$ of eqn (3.115) are per nonet of residues (a triplet from each of three chains); the parameters Δh_a, etc. are per residue. Eqn (3.115) contains four unknown parameters. Taking $\Delta h_u(\frac{2}{3})$ and $\Delta s_u(\frac{2}{3})$ as given from our study of $(Gly-Pro-Pro)_N$ (eqn (3.111)) reduces the number of unknowns to two. Taking two experimental values of T_m for natural collagen from Fig. 3.14 and using eqn (3.114) and (3.115) allows all the parameters to be determined. We obtain[9]

$$\Delta h_i = -4975 \text{ J mol}^{-1},$$
$$\Delta s_i = -13 \text{ J mol}^{-1} \text{ K}^{-1},$$
$$\Delta h_a = -337 \text{ J mol}^{-1}, \quad (3.116)$$
$$\Delta s_a = -1.9 \text{ J mol}^{-1} \text{ K}^{-1}.$$

The solid curve in Fig. 3.14 shows the variation of T_m on α as calculated using the parameters of eqn (3.116) in eqn (3.115) and (3.114). We note from the parameters of eqn (3.116) that imino acids contribute a large negative enthalpy to the triple helix, while the enthalpy contribution of amino acids is almost zero. We also see that the entropy loss for triple-helix formation for imino acids is much greater than for amino acids. Based on steric considerations alone, we would expect the reverse: amino acids have more conformational freedom in the random coil, hence lose a greater amount of entropy when transferred to the ordered triple helix. Presumably the large entropy loss for imino acids is a result of the changes of solvent structure (freeing of water molecules, see section 2.2.13) when imino acids are incorporated in the triple helix.

From the general formula (Gly–x–y), and from the assumption of eqn (3.115), there are three types of triplet Gly–x–y in natural collagen that differ in the percentage of imino acids in the triplet. Corresponding to these three types of triplet are three s values, where

$$s(\alpha) = \exp(-\Delta h_u(\alpha)/RT) \exp(\Delta s_u(\alpha)/R). \quad (3.117)$$

Table 3.4 shows $s(0)$, $s(\frac{1}{3})$, and $s(\frac{2}{3})$ as functions of temperature calculated using eqn (3.117) and the parameters of eqn (3.116). We can use the data of Table 3.4 to construct a partition function for specific-sequence natural collagen, a topic

TABLE 3.4.

Helix-formation constant for adding three identical triplets (one from each chain) to the triple helix of collagen as a function of the fraction α of imino acids per triplet and temperature

T (K)	s ($\alpha=0$)	s ($\alpha=\frac{1}{3}$)	s ($\alpha=\frac{2}{3}$)
0	0.51	4.01	33.50
300	0.48	2.17	10.80
325	0.42	1.29	4.12
350	0.38	0.82	1.81
375	0.35	0.56	0.89

we shall return to in Chapter 6.

While eqn (3.115) represents the simplest assumption we can make to fit the experimental data of Fig. 3.14, it gives, based on the parameters obtained from the model system (Gly–Pro–Pro)$_N$ and two experimental melting temperatures of natural collagen, a first estimate of the nature of the heterogeneity of helix stability $s(\alpha)$ in natural collagen.

3.5. Sequence statistics in natural collagen[12]

From Table 3.4 we see that the imino-acid content of a triplet enormously influences the stability of a triplet in the triple helix. Thus it is important to ascertain whether or not the imino acids tend to occur more or less at random or whether they tend to cluster in the sequence.

In the general formula Gly–x–y there are two positions (x and y), either or both of which can be imino acids. Based on this dichotomy there are four different types of triplet, (a,a), (i,a), etc. The simplest approach to the sequence statistics is to consider each triplet as having two sites, x and y, where an imino acid can 'bind' (thus we use the mathematical analogue of a titration curve). For two independent (independence gives random occurrence) sites per triplet the grand partition function for a molecule is

$$\xi = (1+\gamma)^{2N}, \qquad (3.118)$$

where N is the number of triplets in a single strand, the exponent $2N$ arising since a triplet has two sites for the occurrence of an imino acid. The parameter γ plays the role of λ in the ordinary use of the grand partition function. The quantity that is available experimentally is the average fraction of imino acids per triplet. Then γ is given in terms of α in the standard fashion

$$\frac{3}{2}\alpha = \frac{1}{2N} \frac{\partial \ln \xi}{\partial \ln \gamma}, \qquad (3.119)$$

where the factor $\frac{3}{2}$ is introduced since in a triplet only 2/3 of the positions can be imino acids. Using eqn (3.118) for ξ, eqn (3.119) yields

$$\gamma = \frac{3\alpha}{2 - 3\alpha}. \qquad (3.120)$$

Since ξ per triplet is

$$\xi = (1+\gamma)^2 = 1 + 2\gamma + \gamma^2, \qquad (3.121)$$

we have immediately

$$\begin{aligned} p(a,a) &= 1/\xi, \\ p(a,i) &= p(i,a) = \gamma/\xi, \\ p(i,i) &= \gamma^2/\xi. \end{aligned} \qquad (3.122)$$

To simplify the discussion of runs of triplets we introduce the notation

$$\begin{aligned} p(0) &= p(a,a), \\ p(1) &= p(a,i) + p(i,a), \\ p(2) &= p(i,i), \end{aligned} \qquad (3.123)$$

the index in $p(k)$ indicating the number of imino acids in a triplet regardless of position. Then the *a priori* probabilities of pairs of triplets are given by

$$p(00) = p(0)p(0),$$
$$p(01) = p(0)p(1) = p(10),$$
$$p(11) = p(1)p(1),$$
$$p(02) = p(0)p(2) = p(20), \quad (3.124)$$
$$p(12) = p(1)p(2) = p(21),$$
$$p(22) = p(2)p(2).$$

The analysis of the occurrence of pairs of triplets represents only a first indication of randomness. Systems that show random statistics for pair occurrence need not be random. Thus if the sequence ... 001100110011 ... were analysed for the occurrence of the pairs (01), (10), (11), and (00) we would find exact agreement with the result predicted by the random occurrence of 0 and 1 each occurring with equal probability. A more sensitive test of randomness is the analysis of the frequency of runs of triplets of a given composition. In Chapter 4 we will show that the probability of a sequence of j consecutive independent units in a chain is given by

$$p(k_j) = \{1-p(k)\}p(k)^{j-1}, \quad (3.125)$$

where $p(k)$ is the *a priori* probability that a triplet contains k imino acids regardless of the order, $k = 0$, 1, or 2. The distribution $p(k_j)$ is defined such that

$$\sum_{j=1}^{\infty} p(k_j) = 1. \quad (3.126)$$

The probabilities in eqns (3.122), (3.124), and (3.125) are all given in terms of the parameter γ, which through eqn (3.120) is determined by α, the average fraction of imino acids in a strand.

Using a composite of sequence data having an average value $\alpha = 0.23$ (eqn (3.120) thus gives $\gamma = 0.53$), Table 3.5 shows a comparison of experimental and calculated probabilities of occurrence of triplets (eqn (3.122)) and pairs of triplets (eqn (3.124)). The numbers in Table 3.5 show that as far as the occurrence of different triplets and pairs of triplets is concerned, the occurrence of imino acids in the sequence are random. Fig. 3.15 shows calculated and experimental values of $p(k_j)$. Again there is no significant difference from random occurrence. Thus there is no tendency for imino acids to cluster in the sequence.

TABLE 3.5.

Probabilities of the occurrence of triplets and pairs of triplets in collagen; experimental data are for a composite of sequence data while the calculated values are for random occurrence of imino acids

Probability	Experimental	Calculated
$p(a,a)$	0.475	0.430
$p(a,i)$	0.245	0.226
$p(i,a)$	0.235	0.226
$p(i,i)$	0.105	0.118
$p(00)$	0.178	0.184
$p(10)$	0.205	0.194
$p(01)$	0.188	0.194
$p(11)$	0.228	0.204
$p(02)$	0.045	0.051
$p(20)$	0.035	0.051
$p(21)$	0.051	0.054
$p(22)$	0.010	0.014

3.6. Summary

In this chapter we have shown how the analysis of the association of synthetic single strands of polynucleotides and polypeptides to give respectively double and triplet helices yields thermodynamic parameters that can be used in naturally occurring molecules. The importance of these studies is that the composition, molecular weight, and nature of the equilibria are well characterized. Naturally occurring macromolecules contain too many effects operating simultaneously (as will be

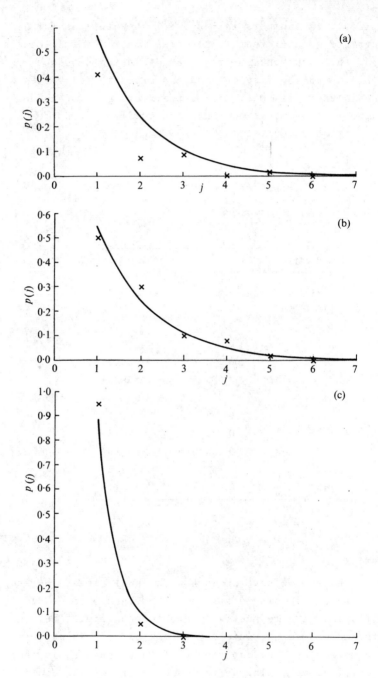

discussed in Chapter 6) to yield a reliable set of parameters from a direct study of such systems. The multiple equilibria in the systems treated in this chapter are very simple, the sums required involving variants of the geometric series (e.g. eqns (2.37) and (3.102)). As chains become longer, there are more structural possibilities and the partition functions required to treat such systems also become more complicated. In the next chapter we present general techniques of evaluating partition functions for linear chain molecules.

Notes and References

1. Eqn (3.4) can be written $[S_1] = \zeta c_0$, which illustrates that ζ plays the role of an activity coefficient.
2. The development of this section follows that given in: Schwarz, M. and Poland, D., *Biopolymers* **13**, 687 (1974).
3. The kind of partition function used in this section for short chains was first used by Schellman to treat the helix-coil transition in polypeptides: Schellman, J.A., *J. phys. Chem.* **62**, 1485 (1958).
4. Applequist, J. and Damle, V., *J. Am. chem. Soc.* **87**, 1450 (1965).
5. See ref. 7 in Chapter 1.
6. Scheffler, I.E., Elson, E.L., and Baldwin, R.L., *J. molec. Biol.* **36**, 291 (1968); **48**, 145 (1970).
7. Fisher, M.E., *J. chem. Phys.* **45**, 1469 (1966).
8. See P&S, Chapter 9.
9. See ref. 2.
10. Kobayashi, Y., Sakai, R., Kakinchi, K., and Isemura, T., *Biopolymers* **9**, 415 (1970).
11. See ref. 18 in Chapter 2.
12. Schwarz, M. and Poland, D., *Biopolymers* **13**, 1873 (1974).

FIG. 3.15. Calculated (solid curve) and experimental (crosses) sequence-length distributions in natural collagen; the calculated curves are for random occurrence of imino acids. (a) Probability of having j triplets in a row each having no imino acids; (b) probability of having j triplets in a row each having one imino acid per triplet; (c) probability of having j triplets in a row each having two imino acids per triplet.

4
LINEAR CHAIN PARTITION FUNCTIONS: TECHNIQUE AND APPLICATIONS

In this chapter we systematically explore techniques for the evaluation of partition functions for linear chain molecules and give examples of what the application of technique tells one about the properties of biological macromolecules.[1] A short review of matrix algebra is given in the Appendix.

4.1. Matrix method

In general a partition function is a sum over states whether the states are quantum-mechanical energy levels, free energy levels, states of binding, or amount of double (or triple) helix. In Chapter 1 I gave the partition function (sum over states of binding) for a polyprotic acid having N independent carboxyl groups. We showed that the partition function was (eqn (1.82))

$$\xi = \sum_{n=0}^{N} \frac{N!}{\overline{N-n}\ !n!} ([H^+]K)^n. \qquad (4.1)$$

We further showed that the sum in eqn (4.1) can be evaluated to give

$$\xi = (1 + [H^+]K)^N. \qquad (4.2)$$

On comparing eqn (4.1) with eqn (4.2) we see that the partition function in eqn (4.1) (sum over states of binding) is evaluated as a product in eqn (4.2). The property of being able to write the partition function as a product is general if the states per unit (states of binding, different conformations) are independent of the states of neighbouring units.

An example is the stacking equilibrium in single-strand polynucleotides. In the single-strand state there is a tendency for the neighbouring bases to stack even without the stabilization of inter-base hydrogen bonds. To first approximation the extent of stacking is independent of chain length, indicating that the equilibrium

LINEAR CHAIN PARTITION FUNCTIONS:

(unstacked neighbouring bases) ⇌ (stacked neighbouring bases) (4.3)

is independent of the states of stacking of neighbouring bases. For a model we assume there are two states per residue, stacked and unstacked, with equilibrium constant of interconversion s (eqn (4.3)). Then the partition function for a chain of N bases is given by

$$q = \sum_{n=0}^{N} \frac{N!}{(N-n)!n!} s^n = (1+s)^N, \qquad (4.4)$$

where n is the number of bases stacked. The fraction of stacked bases is given by

$$\vartheta = \frac{\langle n \rangle}{N} = \frac{1}{N} \frac{\partial \ln q}{\partial \ln s} = \frac{s}{1+s}. \qquad (4.5)$$

Note that the equilibrium reduces to that of the simple two-level system discussed in section 1.1 with $\exp(-\Delta\varepsilon/RT)$ replaced by s. An experimental test of independent states is that, as shown in eqn (4.5), an average quantity is independent of chain length. As in eqn (3.66) we can express s in the form

$$s = \exp\{\Delta h_u (T-T_0)/RTT_0\}, \qquad (4.6)$$

where in this case T_0 and Δh_u are determined experimentally from the relations

$$\vartheta(s=1) = \vartheta(T_0) = \tfrac{1}{2}, \qquad (4.7a)$$

$$\left(\frac{\partial \ln \vartheta}{\partial (1/T)}\right)_{\vartheta=\tfrac{1}{2}} = -\frac{\Delta h_u}{2R}. \qquad (4.7b)$$

For poly(A) at pH = 7 (poly(A) only associates to give double helix as discussed in section 3.1.5 at a pH of about 4) we obtain[2]

$$\Delta h_u = -27.2 \text{ kJ mol}^{-1},$$
$$\Delta s_u = -90 \text{ J mol}^{-1} \text{ K}^{-1}, \qquad (4.8)$$
$$T_0 = 300 \text{ K}.$$

In terms of the parameters introduced in section 3.2.3 the s for stacking can be interpreted as $s = \tau/u$. The τ for stacking in single strands need not be the same τ appropriate to the double helix, but we see that the parameters of eqn (4.8) are of the same magnitude as those of eqn (3.72) (for the formation of double helix in poly(A–T). Since for the double helix $s = t\tau/u^2$, t representing the effect of base–base hydrogen bonding, we see that t makes only a small contribution to s.

In general if z_i are the statistical weights for the independent states of a unit in a macromolecule then the partition function for the whole molecule is generated by taking the product of the sum of the statistical weights of a unit

$$q = \left(\sum_i z_i\right)^N, \qquad (4.9)$$

with the probability of the ith state independent of chain length

$$p_i = z_i / \sum_i z_i. \qquad (4.10)$$

We now show that the partition function for the case where the states of a unit are not independent of the states of neighbouring units can also be generated as a product but not simply as the product of the linear sum of the statistical weights per unit.

Suppose that the statistical weights for states of a unit depend on the state of the neighbouring unit such that, for example, if there are two states a and b per unit, statistical weights such as z_{ab}, z_{bb}, etc. are required. We can construct a table giving the statistical weights of a given unit in terms of the states of that unit (ith unit) and in terms of the states of the following unit (($i+i$)th unit) as follows

i \ $i+1$	a	b
a	z_{aa}	z_{ab}
b	z_{ba}	z_{bb}

(4.11)

TECHNIQUE AND APPLICATIONS   231

The row indices are the states of the ith unit while the column indices are the states of the $(i+1)$th unit; the entries in the table give the statistical weights appropriate to the ith unit being in a given state followed by the $(i+1)$th unit being in a given state. A particular sequence of states such as

$$\ldots a\ a\ b\ a\ b\ b \ldots$$

would have the net statistical weight

$$\ldots z_{aa}\ z_{ab}\ z_{ba}\ z_{ab}\ z_{bb}\ \ldots \qquad (4.12)$$

The partition function is the sum over all the net statistical weights corresponding to all possible combinations of a and b (if there are N units with two states per unit there are 2^N different sequences of as and bs). We observe that the sum over all net statistical weights as given in eqn (4.12) is generated by the matrix product

$$q = \mathbf{e}\mathbf{W}^N \mathbf{e}^+, \qquad (4.13)$$

where N is the number of units in the chain and $\mathbf{W}$ is the matrix of coefficients in (4.11)

$$\mathbf{W} = \begin{pmatrix} z_{aa} & z_{ab} \\ z_{ba} & z_{bb} \end{pmatrix} \qquad (4.14)$$

Since the partition function is a scalar (a positive number) and a matrix raised to the Nth power generates a new matrix, it is clear that the partition function cannot be simply equated with a matrix product. Thus the left end of the chain is represented by the row vector $\mathbf{e}$ and the right end of the chain is represented by a column vector $\mathbf{e}^+$. Thus eqn (4.13) has the form

$$\text{(row vector)(matrix)(column vector)} = \text{scalar.} \qquad (4.15)$$

The end vectors $\mathbf{e}$ and $\mathbf{e}^+$ can also reflect the fact that the statistical weights of a unit at the physical end of the chain might be different from units in the interior.

To give a concrete example we treat the model of eqn (4.1) of a polyprotic acid having N carboxyl groups, except that now we do not consider the state of ionization of the groups to be independent. The general correlation table of (4.11) becomes for this example

$$
\begin{array}{c|cc}
 i \diagdown {}^{i+1} & \text{COOH} & \text{COO}^- \\
\hline
\text{COOH} & [\text{H}^+]K & [\text{H}^+]K \\
\text{COO}^- & 1 & y
\end{array}
\qquad (4.16)
$$

where as in eqn (4.1) the ionized state COO$^-$ (reference state) is assigned a factor unity and the bound state COOH a factor $[\text{H}^+]K$ (where $K = 10^{+pK}$ is the association constant) with the new feature that if a COO$^-$ is followed in the chain by a COO$^-$ a factor y is assigned. Taking $y < 1$ reflects the fact that two neighbouring charged groups repel each other thus introducing a positive free energy (largely an electrostatic energy of repulsion). In Chapter 5 I will discuss cooperative titrations in detail and will give models for y; here we simply want to treat the mathematical technique for introducing such a feature.

To construct the end vectors we note:

State of second unit	Statistical weights of first unit
COOH	$[\text{H}^+]K + 1$
COO$^-$	$[\text{H}^+]K + y$

(4.17)

State of Nth unit	Statistical weight of Nth unit
COOH	$[\text{H}^+]K$
COO$^-$	1

and that

$$[1 \quad 1] \begin{bmatrix} [H^+]K & [H^+]K \\ 1 & y \end{bmatrix} = \left[([H^+]K+1, \; [H^+]K+y \right],$$

$$\begin{bmatrix} [H^+]K & [H^+]K \\ 1 & y \end{bmatrix} \begin{bmatrix} 1 \\ 0 \end{bmatrix} = \begin{bmatrix} [H^+]K \\ 1 \end{bmatrix}$$

(4.18)

Thus the partition functions for a chain having N carboxyl groups with an interaction statistical weight y between neighbouring charged groups is

$$\xi = [1 \quad 1] \begin{bmatrix} [H^+]K & [H^+]K \\ 1 & y \end{bmatrix}^N \begin{bmatrix} 1 \\ 0 \end{bmatrix}$$

(4.19)

For $N = 3$ there are $2^3 = 8$ different states of binding. Letting (-) and (0) respectively represent charged and uncharged groups and taking $[H^+]K = x$, then the partition function is (illustrating the charge state for each term)

$$\begin{array}{cccccccc}
& 1 & + & x & + & x & + & x & + \\
& (0\ 0\ 0) & & (0\ 0\ -) & & (0\ -\ 0) & & (-\ 0\ 0) & \\
+ & x^2 y & + & x^2 & + & x^2 y & + & x^3 y^2 & \\
& (0\ -\ -) & & (-\ 0\ -) & & (-\ -\ 0) & & (-\ -\ -) &
\end{array}$$

(4.20)

$$= [1 \quad 1] \begin{bmatrix} x & x \\ 1 & y \end{bmatrix}^3 \begin{bmatrix} 1 \\ 0 \end{bmatrix}.$$

While we could generate the partition function by actually performing the matrix multiplication (for many problems this is in fact the simplest procedure) we can obtain a still simpler form for the partition function. In matrix algebra (see the Appendix) it is shown that we can find matrices $\mathbf{T}$ and $\mathbf{T}^{-1}$ such that[3]

$$W = T\Lambda T^{-1},$$

$$TT^{-1} = I = \begin{bmatrix} 1 & 0 \\ 0 & 1 \end{bmatrix},$$

$$\Lambda = \begin{bmatrix} \lambda_1 & 0 \\ 0 & \lambda_2 \end{bmatrix}, \qquad (4.21)$$

where I is the identity matrix and Λ is a diagonal matrix. Using the above, eqn (4.13) becomes

$$q = e(T\Lambda T^{-1})^N e^+ = eT\Lambda^N T^{-1} e^+. \qquad (4.22)$$

The utility of introducing the diagonal matrix Λ is that

$$\Lambda^N = \begin{bmatrix} \lambda_1 & 0 \\ 0 & \lambda_2 \end{bmatrix}^N = \begin{bmatrix} \lambda_1^N & 0 \\ 0 & \lambda_2^N \end{bmatrix}. \qquad (4.23)$$

Thus q of eqn (4.22) becomes

$$q = c_1 \lambda_1^N + c_2 \lambda_2^N, \qquad (4.24)$$

where c_1 and c_2 are factors arising from the T and T^{-1} matrices and the end vectors. A detailed recipe for calculating the cs from matrix algebra is given elsewhere;[4] and alternative derivation not using matrix algebra is given in the next section. The λs (eigenvalues or characteristic values of W) are determined from the determinant

$$|W - \lambda I| = 0. \qquad (4.25)$$

For the general matrix of eqn (4.14), eqn (4.25) is

$$\begin{vmatrix} z_{aa} - \lambda & z_{ab} \\ z_{ba} & z_{bb} - \lambda \end{vmatrix} = (z_{aa} - \lambda)(z_{bb} - \lambda) - z_{ab} z_{ba} = 0 \qquad (4.26)$$

TECHNIQUE AND APPLICATIONS

or

$$\lambda^2 - \lambda(z_{aa}+z_{bb}) + z_{aa}z_{bb} - z_{ab}z_{ba} = 0. \qquad (4.27)$$

Solving the quadratic equation gives

$$\lambda_1 = \tfrac{1}{2}\{-B + \sqrt{(B^2-4C)}\}, \quad \lambda_2 = \tfrac{1}{2}\{-B - \sqrt{(B^2-4C)}\}; \qquad (4.28)$$

$$B = -(z_{aa}+z_{bb}), \qquad C = z_{aa}z_{bb} - z_{ab}z_{ba}.$$

Ordering the eigenvalues λ such that λ_1 is the largest we can write eqn (4.24) as

$$q \quad c_1 \lambda_1^N \left\{1 + \frac{c_2}{c_1}\left(\frac{\lambda_2}{\lambda_1}\right)^N\right\}.$$

As N becomes large $(\lambda_2/\lambda_1)^N$ rapidly goes to zero (since, by choice $\lambda_2 < \lambda_1$). Then we have

$$\frac{1}{N} \ln q = \frac{1}{N} \ln c_1 + \ln \lambda_1. \qquad (4.29)$$

As N goes to infinity we have

$$\lim_{N \to \infty} \frac{1}{N} \ln q = \ln \lambda_1. \qquad (4.30)$$

Thus very long chains can be described solely in terms of λ_1, the largest eigenvalue of the matrix **W**.

Returning to the model of a chain with N carboxyl groups having a repulsive interaction represented by the factor y between neighbouring charged groups we have (using the specific matrix of eqn (4.1) in eqns (4.25) or (4.26) with $x = [H^+]K$)

$$\lambda_1 = \tfrac{1}{2}\left[(x+y) + \sqrt{\{(x+y)^2 + 4x(1-y)\}}\right],$$

$$\lambda_2 = \tfrac{1}{2}\left[(x+y) - \sqrt{\{(x+y)^2 + 4x(1-y)\}}\right]. \qquad (4.31)$$

In general the fraction of bound protons is given by (see eqn (1.64))

$$\vartheta = \frac{1}{N}\frac{\partial \ln q}{\partial \ln x}.$$

For large N using eqn (4.30) we have

$$\vartheta = \frac{\partial \ln \lambda_1}{\partial \ln x}, \qquad (4.32)$$

which gives an equation for the titration curve; if K is known for an isolated COOH group then a single measurement of ϑ at known pH allows y to be calculated. In general because of eqn (4.30) the average properties of large chains are given in terms solely of the largest eigenvalue, λ_1.

If there is no interaction between neighbouring COO$^-$ groups ($y = 1$) then eqn (4.31) gives

$$\lambda_1 = x + 1,$$
$$\lambda_2 = 0. \qquad (4.33)$$

We also find that $c_1 = 1$ and $c_2 = 0$. Hence eqn (4.24) becomes in the special case of $y = 1$

$$q = (1+x)^N \qquad (4.34)$$

which is eqn (4.2) (or eqn (4.1)).

If there are more than two states per unit with nearest-neighbour interactions eqn (4.16) is easily generalized, e.g.

$i+1$ \ i	a	b	c
a	z_{aa}	z_{ab}	z_{ac}
b	z_{ba}	z_{bb}	z_{bc}
c	z_{ca}	z_{cb}	z_{cc}

(4.35)

If units interact with more than just nearest neighbours, then the table of (4.16) is expanded to include all possibilities. For example, for the correlation of three consecutive units with two states per unit, the table required is

	$i+2$	a	b	a	b
i	$i+1$	a	a	b	b
a	a	z_{aaa}	z_{aab}	0	0
a	b	0	0	z_{aba}	z_{abb}
b	a	z_{baa}	z_{abb}	0	0
b	b	0	0	z_{bba}	z_{bbb}

(4.36)

The zeros in (4.36) are for entries that do not make sense (i.e. where the $(i+1)$th unit is simultaneously a and b). Note that three consecutive units with two states per unit have $2^3 = 8$ possible states, the number of non-zero entries in (4.36). In general to correlate μ units with ν states per unit requires a matrix of at least the size

$$\text{matrix size} = \nu^{\mu-1} \times \nu^{\mu-1}. \tag{4.37}$$

The actual matrix size for a specific model can be smaller than the size given in eqn (4.37) if the correlation of μ consecutive units is not required for all the states.

If $\rho = \nu^{\mu-1}$ is the matrix size, then in general the partition function (as in eqn (4.24)) will be given by

$$q = \sum_{i=1}^{\rho} c_i \lambda_i^N, \tag{4.38}$$

there being ρ eigenvalues for a $\rho \times \rho$ matrix (i.e. eqn (4.25) generates a ρth-order polynomial in λ from a $\rho \times \rho$ matrix).

4.2. Sequence conditional probabilities[5]

The matrix method described in the last section gives the partition function (eqn (4.38)) exactly for a given model for finite chains. We will see in subsequent sections that the finite length of biological macromolecules can have a pronounced influence on the conformational equilibria. Thus the matrix technique is very important since it correctly includes the influence of finite chain length. For very long chains we have seen in eqn (4.30) that knowledge of the largest eigenvalue is sufficient to describe the properties of the chain. In this section I shall develop a very simple technique to obtain the largest eigenvalue, thus giving the properties of very long chains.

Consider a model where each unit in the chain can exist in two states, a and b. Every specific combination of as and bs such as

$$\ldots a\,a\,a\,b\,b\,b\,b\,a\,b\,b\,a\,a\,b\,b\,b \ldots \qquad (4.39)$$

can be viewed as consisting of alternating sequences of a and b states. We let $u(a_j)$ and $u(b_j)$ respectively represent the statistical weights of sequences of a's and b's j units long; in this model the statistical weight of a sequence of as and bs depends on the length j of the sequence but not on the length of neighbouring sequences. Consider the two configurations of states for the chain

$$\begin{aligned}&\text{(I)} \quad a\ldots i\ldots ab\ldots j\ldots bx\ldots\ldots x,\\ &\text{(II)} \quad a\ldots i\ldots ax\ldots j{+}k\ldots\ldots\ldots x,\end{aligned} \qquad (4.40)$$

where the notation a...i...a represents a sequence of a states i units long and x represents the case where a unit can be either a or b. The ratio of the partition functions for chain configurations (I) and (II) gives the conditional probability that given a sequence of a's i units long, a sequence of b's j units long follows

$$P(a_i|b_j) = q_I\,q_{II}, \qquad (4.41)$$

since in both configurations the sequence of a's is fixed, it being stipulated in (I) that a sequence of j b's follows, configuration (II) allowing all length sequences of b's. Thus

$$P(a_i|b_j) = \frac{\text{term}\,(a_i|b_j)}{\sum_j \text{term}\,(a_i|b_j)}, \qquad (4.42)$$

TECHNIQUE AND APPLICATIONS

which is seen to be identical with our basic relation for constructing probabilities (eqn (1.77)).

Using the matrix technique, the ratio of the qs is given by

$$P(a_i|b_j) = \frac{q_I}{q_{II}} \frac{u(a_i)u(b_j)\mathbf{e}_a \mathbf{W}^k \mathbf{e}_x^+}{u(a_i)\mathbf{e}_a \mathbf{W}^{j+k} \mathbf{e}_x^+}, \qquad (4.43)$$

where the contributions of the sequences of x's in (4.40) have been written as matrix products. The end vectors $\mathbf{e}_b$, etc. reflect the nature of the boundary; other than indexing them, we will not need to specify the nature of the end vectors. From eqn (4.29) we have, for example,

$$\lim_{N \to \infty} \mathbf{e}_a \mathbf{W}^N \mathbf{e}_b^+ = c_{ab}\lambda_1^N, \qquad (4.44)$$

while from the general technique for constructing the factors c we have

$$c_{ab} = L_a R_b. \qquad (4.45)$$

That is c_{ab} factors into contributions from the left end and right end of the chain. On letting k go to infinity in eqn (4.43), using eqns (4.44) and (4.45) we find

$$P(a_i|b_j) = (L_b/L_a)u(b_j)\lambda_1^{-j}. \qquad (4.46)$$

In a similar manner (reversing the roles of a and b in (4.40)) we find

$$P(b_i|a_j) = (L_a/L_b)u(a_j)\lambda_1^{-j}. \qquad (4.47)$$

The conditional probabilities have the property

$$\sum_j P(a_i|b_j) = 1,$$

$$\qquad (4.48)$$

$$\sum_j P(b_i|a_j) = 1,$$

i.e. given, for example, a sequence of i as, it must be followed by a sequence of bs of some length. Using eqns (4.46) and (4.47) in eqn (4.48) yields

$$(L_a/L_b) \sum_{j=1}^{\infty} u(a_j)\lambda_1^{-j} = 1,$$
(4.49)

$$(L_b/L_a) \sum_{j=1}^{\infty} u(b_j)\lambda_1^{-j} = 1.$$

Mutual mutliplication of the above expressions gives

$$U_a(\lambda) U_b(\lambda) = 1, \qquad (4.50)$$

where

$$U_a(\lambda) = \sum_{j=1}^{\infty} u(a_j)/\lambda^j,$$
$$U_b(\lambda) = \sum_{j=1}^{\infty} u(b_j)/\lambda^j.$$
(4.51)

The values of λ that satisfy eqn (4.50) are the eigenvalues of the matrix **W**; thus eqn (4.50) is the secular equation of **W**. Note that we have not had to specify the nature of **W**. The quantities in eqn (4.51) are known as generating functions.[6] Thus for any model that requires general statistical weights for alternating sequences of states that depend in an arbitrary fashion on the size of the sequence (but not on the length of neighbouring sequences), the solutions of eqn (4.50) give all the eigenvalues and in particular the largest. Although the partition function is always a real number, the roots of eqn (4.50) can be complex except for the largest root which is always real.

Lifson[6] has shown that if we have more than two states per unit, the analogue of eqn (4.50) is

TECHNIQUE AND APPLICATIONS   241

$$|M - I| = 0, \quad (4.52)$$

where

$$M = \begin{bmatrix} 0 & U_{ab} & U_{ac} \\ U_{ba} & 0 & U_{bc} \\ U_{ca} & U_{cb} & 0 \end{bmatrix} \quad (4.53)$$

where U_{ab}, etc. are the generating functions for a particular type of sequence followed by another type (including at most a nearest-neighbour interaction for the boundary between the two types of states). Note that in eqn (4.53) the elements for b followed by b, etc. are missing since these do not make sense (a sequence of bs followed by a sequence of bs is just a larger sequence of bs).

We now show that for a very useful class of models the partition function for finite N can be calculated exactly solely in terms of the generating functions. Consider the case where the statistical weights for sequences of one state (here taken as b) can be written as a product of factors

$$u(b_j) = u_b^{\,j} \quad (4.54)$$

and where the ends of the chain can be treated as if they were bordered by b states (i.e. there is no special nearest-neighbour interaction when an a unit is bordered by a b unit, hence the units on the end can be considered to be always bordered by bs for simplicity). Then in eqn (4.38) the c factors can be written specifically as (using eqn (4.45))

$$c = L_b R_b. \quad (4.55)$$

To determine $L_b R_b$ consider the two chain configurations

$$\begin{aligned}&(I) \quad ax\ldots\ldots n\text{-}1\ldots xbx\ldots m\ldots xa, \\ &(II) \quad ax\ldots\ldots\ldots n\text{+}m\text{-}1\ldots\ldots xa.\end{aligned} \quad (4.56)$$

Following our treatment of (4.40) we have

$$P(a|n|b) = q_I/q_{II} = (L_b/L_a)u_b\lambda_1^{-n}e_a W^{n-1}e_b^+, \qquad (4.57)$$

where $P(a|n|b)$ is the conditional probability that given an a unit, a unit b follows n units away. On letting n become infinite we have

$$\lim_{N \to \infty} P(a|n|b) = p_b = L_b R_b u_b \lambda_1^{-1}, \qquad (4.58)$$

where p_b is the *a priori* probability that a state is b. For infinite chains

$$p_b = \frac{\partial \ln \lambda_1}{\partial \ln u_b} = \frac{u_b}{\lambda_1}\frac{\partial \lambda_1}{\partial u_b}. \qquad (4.59)$$

Comparing eqn (4.58) and (4.59) we have

$$c_{bb} = \frac{\partial \lambda_1}{\partial u_b}. \qquad (4.60)$$

Using eqn (4.54) for $u(b_j)$, the generating function $U_b(\lambda)$ (eqn (4.51)) is (the sum is the geometric series)

$$U_b(\lambda) = u_b/(\lambda - u_b). \qquad (4.61)$$

Using eqn (4.61) for $U_b(\lambda)$ and implicitly differentiating eqn (4.50) gives

$$c_{bb} = \lambda_1 / \left\{1 - \left(\frac{\partial U_a}{\partial \lambda}\right)_{\lambda_1}\right\}. \qquad (4.62)$$

Since the factors c_i in eqn (4.48) depend only on λ_i, eqn (4.62) yields a general relation for the c_i, λ_1 being replaced by λ_i. Using eqns (4.62) and (4.38) (for the model of eqn (4.54) with $u(a_j)$ general) we have the exact partition function as a function of N

$$q = \sum_{i=1}^{\rho} \lambda_i^{N+1} / \left\{1 - \left(\frac{\partial U_a}{\partial \lambda}\right)_{\lambda_i}\right\}. \qquad (4.63)$$

The derivative is

TECHNIQUE AND APPLICATIONS   243

$$\frac{\partial U_a}{\partial \lambda} = -\lambda^{-1} \sum_{j=1}^{\infty} j \frac{u(a_j)}{\lambda^j} \qquad (4.64)$$

with the λ_i the roots of

$$\left(\frac{u_b}{\lambda - u_b}\right) \sum_{j=1}^{\infty} u(a_j)/\lambda^j = 1. \qquad (4.65)$$

Note that although we made use of limiting forms (large N) in eqn (4.58) the c_i and λ_i are not functions of N and hence eqn (4.63) gives the N dependence exactly for all N.

Of great importance for the physical understanding of multiple equilibria in polymers is the probability distribution for sequence lengths. To derive an expression for the *a priori* probability of a sequence of states of a given length consider the two chain configurations

$$\begin{array}{l}(I) \quad x\ldots.m\ldots.xa\ldots.j\ldots.ax\ldots.m\ldots x, \\ (II) \quad x\ldots\ldots\ldots\ldots.2m+j\ldots\ldots\ldots.x.\end{array} \qquad (4.66)$$

The *a priori* probability $p(a_j)$ of a sequence of as j units long is then given simply by the ratio of the partition functions for the configurations in (4.66)

$$p(a_j) = q_I/q_{II} = c_{aa} u(a_j) \lambda_1^{-j} \qquad (4.67)$$

where eqn (4.67) gives the long-chain limit. Eqn (4.67) gives the probability that a sequence of as occurs at a given position in the chain; since in an infinite chain all positions are equivalent, it thus gives the probability *a priori* of the occurrence of such a sequence appearing anywhere as compared to all other possible lengths and types of sequence. The probability distribution can be normalized solely in terms of sequences of one type as

$$p(a_j) = \frac{u(a_j)/\lambda_1^j}{\sum_{j=1}^{\infty} u(a_j)/\lambda_1^j}. \qquad (4.68)$$

Another probability distribution that proves informative is the probability that a unit appears in a sequence of a given length. We have that this probability is proportional to the *a priori* probability of a sequence of a given length times the number of units in the sequence. Denoting this probability by $P(a_j)$, we have

$$P(a_j) \sim j p(a_j), \qquad (4.69)$$

or using eqn (4.67) and normalizing,

$$P(a_j) = \frac{j u(a_j)/\lambda_1^j}{\sum_{j=1}^{\infty} j u(a_j)/\lambda_1^j}. \qquad (4.70)$$

As we will see in section 4.5, the distributions $p(a_j)$ and $P(a_j)$ have quite different properties. Thus here we stress what they represent:

$$p(a_j) = \begin{pmatrix} a\ priori \text{ probability that, given a sequence} \\ \text{of } a\text{'s, it will contain } j \text{ units} \end{pmatrix}$$

$$P(a_j) = \begin{pmatrix} a\ priori \text{ probability that, given a unit } a, \\ \text{it will be in a sequence of } j \text{ } a \text{ units} \end{pmatrix}. \qquad (4.71)$$

4.3. Combinatorial formulation

Eqn (4.1) is an example of the combinatorial formulation of a partition function. For two independent states per unit eqn (4.1) can be written

$$q = \sum_{n=0}^{N} g(N,n) z_a^n z_b^{N-n}, \qquad (4.72)$$

where

$$g(N,n) = \begin{pmatrix} \text{number of distinct ways } n \text{ states } a \text{ can be} \\ \text{arranged on a chain of } N \text{ units} \end{pmatrix}$$

$$= \frac{N!}{(n-n)!n!}. \qquad (4.73)$$

The quantity $g(N,n)$, the combinatorial factor, gives the number of ways a given number of units a can be placed on a chain of N units when the states are independent. Other examples of the combinatorial formulation of the partition function are eqn (3.37) and eqn (3.102), where the partition function was given in terms of the number of ways of placing a helical sequence of a given length on chains each of N units. While in general the formulation of the partition function in terms of explicit combinatorial factors is difficult for complicated systems, there are a few examples that provide insight into multiple equilibria in biopolymers.

For the model of a chain with two states per unit, a and b, where the statistical weights depend on the state of a unit and the state of the neighbouring unit let

N_a = number of a states,

N_b = number of b states,

$N_a + N_b = N$, (4.74)

N_s = number of sequences of as = number of sequences of bs.

Note that the last equation arises from the fact that sequences of as and bs must alternate, hence there are the same number of each (ignoring the ends of the molecule). The number of distinct ways N_s partitions can be placed among N_a units is

$$\frac{N_a!}{(N_a - N_s)! N_s!},$$

with a similar expression for b units. Thus the total number of distinct ways we can arrange N_a a units and N_b b units into N_s sequences is

$$g(N, N_a, N_s) = \frac{N_a!}{(N_a - N_s)! N_s!} \frac{(N - N_a)!}{(N - N_a - N_s)! N_s!}. \tag{4.75}$$

Using eqn (4.75) the partition function is[7]

$$q = \sum_{N_a} \sum_{N_s} g(N, N_a, N_s) \left(z_{ab}\right)^{N_s} \left(z_{ba}\right)^{N_s} \left(z_{aa}\right)^{N_a - N_s} \left(z_{bb}\right)^{N - N_a - N_s} \quad (4.76)$$

$$= z_{bb}^{N} \sum_{N_a} \sum_{N_s} g(N, N_a, N_s) \left[\frac{z_{ab} z_{ba}}{z_{aa} z_{bb}}\right]^{N_s} \left(\frac{z_{aa}}{z_{bb}}\right)^{N_a}$$

which is an alternative form for the partition function given by eqns (4.13) and (4.14). The sums in eqn (4.76) cannot be put in closed form and hence the maximum-term method must be used to find the values of N_a and N_s that maximize the general term. Since the maximum-term method is only accurate in the limit as N becomes infinite all effect of the ends (finiteness) of the molecule is lost.

For the model treated by eqn (4.50) (general statistical weights $u(a_j)$ and $u(b_j)$ for sequences of as and bs) the combinatorial formulation of the partition function is[8]

$$q = \sum_{N_s} \sum_{n_j} \sum_{m_j} \prod_j \frac{N_s!}{n_j!} u(a_j)^{n_j} \prod_j \frac{N_s!}{m_j!} u(b_j)^{m_j}, \quad (4.77)$$

where N_s is defined in eqn (4.74) and n_j and m_j are respectively the number of sequences of as and bs j units long. The sums in eqn (4.77) have the constraints

$$\Sigma n_j = N_s,$$
$$\Sigma m_j = N_s, \quad (4.78)$$
$$\Sigma j n_j + \Sigma j m_j = N.$$

Again the maximum-term method must be used where the constraints of eqn (4.78) are incorporated using Lagrange undetermined multipliers.

4.4. Simple model for polyethylene[9]

I now want to illustrate the technique we have discussed with

TECHNIQUE AND APPLICATIONS   247

an oversimplified model for conformational equilibria in polyethylene which none the less retains the most important features of the system. To first approximation polyethylene, $CH_2(CH_2)_N CH_3$, can be thought of as a linear array of bonds each of which has barriers of internal rotation similar to ethane. It is known that in molecules like butane,

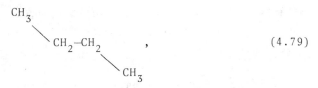

(4.79)

internal rotation about the central bond is similar to that in ethane (three minima) but that the minima are not equal in energy. The three minima correspond to two *gauche* states and the *trans* state (illustrated in Fig. 2.5, p. 124) which we denoted by G^-, T, and G^+. It is found that the two *gauche* states are higher in energy than the *trans* state (in the *gauche* state the two methyl groups are brought very close together, bumping slightly) by about 2.1 kJ mol^{-1}; the potential for internal rotation for butane is illustrated by the dashed curve in Fig. 2.5. Taking the *trans* state as the zero of potential energy we assign the statistical weights

$$z_{G^+} = z_{G^-} = \exp(-\Delta\varepsilon/RT),$$
$$z_T = 1,$$

(4.80)

where $\Delta\varepsilon \simeq 2.1$ kJ mol^{-1} (thus at room temperature $z_{G^+} \simeq 1/e$). If the states of internal rotation for the N bonds about which rotation produces a new carbon conformation (excluding rotation of the end methyl groups which just rotates hydrogens) in $CH_3(CH_2)_N CH_3$ were independent, then the partition function would be

$$q = (z_{G^-} + z_T + z_{G^+})^N.$$

(4.81)

Some of the important probabilities for the system derived from eqn (4.81) are

$$P_{G^+} = P_{G^-} = z_{G^+}/\lambda_1,$$

$$P_T = z_T/\lambda_1 \qquad (4.82)$$

$$\lambda_1 = z_{G^-} + z_T + z_{G^+}.$$

Using space-filling molecular models we find that the states of internal rotation of two neighbouring bonds are not, in fact, independent. In particular we find that if the successive states of two units are G^+G^- of G^-G^+ there is extreme steric hindrance. Thus to first approximation we take the statistical weights for pairs of units as

$$z_{G^+G^-} = z_{G^-G^+} = 0 \qquad (4.83)$$

and treat the other states as independent. The nearest-neighbour correlation table that introduces the effect of eqn (4.83) is

i \ $i+1$	G^-	T	G^+
G^-	z_{G^-}	z_{G^-}	0
T	z_T	z_T	z_T
G^+	0	z_{G^+}	z_{G^+}

$$(4.84)$$

where the entry in the table gives the statistical weight of the appropriate state of the ith unit in terms of the states of the $(i+1)$th unit. Taking

$$z_T = 1, \quad z_{G^+} = z_{G^-} = \exp(-\Delta\varepsilon/RT) = x. \qquad (4.85)$$

The partition function using the matrix given in eqn (4.84) is

$$q = [1 \; 1 \; 1] \begin{bmatrix} x & x & 0 \\ 1 & 1 & 1 \\ 0 & x & x \end{bmatrix}^N \begin{bmatrix} 0 \\ 1 \\ 0 \end{bmatrix} \qquad (4.86)$$

where the end vectors are constructed as follows. The first

unit ('left end') can be either G^-, T, or G^+. The last unit ('right end') is treated as if it is followed by a *trans* state; this is simply a convenient choice, valid since the only nearest-neighbour correlation involves *gauche* states (thus the last unit can be G^-, T, or G^+, with statistical weights assigned as if the next unit is *trans*).

The eigenvalues of the matrix in eqn (4.86) are determined from

$$\begin{vmatrix} x-\lambda & x & 0 \\ 1 & 1-\lambda & 1 \\ 0 & x & x-\lambda \end{vmatrix} = (x-\lambda)\{(1-\lambda)(x-\lambda)-x\} - x(x-\lambda)$$

$$= (x-\lambda)\{(1-\lambda)(x-\lambda)-2x\} = 0.$$
(4.87)

One solution of eqn (4.87) is $\lambda = x$. The other solutions are obtained from

$$(1-\lambda)(x-\lambda) - 2x = 0.$$
(4.88)

The largest root of eqn (4.88) is

$$\lambda_1 = \tfrac{1}{2}[(1+x) + \{(1+x)^2 + 4x\}^{\frac{1}{2}}].$$
(4.89)

As we have seen many times, if there is a one-to-one correspondence between a statistical weight and a particular state, the logarithmic derivative of the partition function with respect to the statistical weight gives the average number of the corresponding states. Since x is assigned both to G^+ and G^- we have (in the limit of large N where the partition function is given by eqn (4.29))

$$p_{G^+} + p_{G^-} = 1 - p_T = \frac{\partial \ln \lambda_1}{\partial \ln x}.$$
(4.90)

The probability of sequences of *trans* and *gauche* states is given from eqn (4.68) (in the limit as N goes to infinity)

$$p(T_j) = \frac{(1/\lambda_1)^j}{\sum_{j=1}^{\infty} (1/\lambda_1)^j} = (\lambda_1 - 1)(1/\lambda_1)^j,$$

$$p(G^+{}_j) = p(G^-{}_j) = \frac{(x/\lambda_1)^j}{\sum_{j=1}^{\infty} (x/\lambda_1)^j} = \frac{(\lambda_1 - x)}{x} \left(\frac{x}{\lambda_1}\right)^j,$$

(4.91)

while the probability that a unit is in a sequence of like states j units long is given by

$$p(T_j) = \frac{j(1/\lambda_1)^j}{\sum_{j=1}^{\infty} j(1/\lambda_1)^j} = (1 - 1/\lambda_1)^2 j (1/\lambda_1)^{j-1},$$

$$p(G^+{}_j) = p(G^-{}_j) = \frac{j(x/\lambda_1)^j}{\sum_{j=1}^{\infty} j(x/\lambda_1)^j} = (1 - x/\lambda_1)^2 j (x/\lambda_1)^{j-1}.$$

(4.92)

It is clear that λ_1, the average partition function per unit in long chains, must be larger than any of the component statistical weights (the lowest value λ_1 could have would be $\lambda_1 = 1$ representing the all *trans* state; mixing the *gauche* states increases λ_1, though not to the value $\lambda_1 = 1 + 2x$ which would apply (eqn (4.81)) if the units were independent. Thus $1/\lambda_1$ and x/λ_1 are less than unity and $p(T_j)$ and $p(G^+{}_j)$ are monotonically decreasing functions of j, i.e. the sequence length $j = 1$ is most probable. However, $P(T_j)$ and $P(G^+{}_j)$ can have a maximum at a value larger than unity (see Figs. 1.3 and 1.5 and the discussion around eqn (1.121)).

Because of the condition imposed by eqn (4.83) G^+ and G^- states cannot mix. Thus any configuration of states has the property that sequences of *trans* states alternate with pure runs of G^+ of G^- states

$$\ldots G^+ \; G^+ \; T \; T \; T \; T \; G^- \; T \; T \; G^- \; G^- \; G^- \; T \; T \; T \; G^+ \; G^+ \; T \; T \; T \ldots, \quad (4.93)$$

where the probability of G^+ and G^- sequences are equal (both

TECHNIQUE AND APPLICATIONS   251

G^+ and G^- states have the same statistical weight x). Since any sequence of units in the same conformational states generates a helix (the all *trans* sequence generates the planar zigzag conformation) conformations such as (4.93) represent alternating runs of short helices.

Table 4.1 gives the various probabilities we have discussed for our simple polyethylene model at several temperatures for both the case where the condition of eqn (4.83) is included and where the three conformational states per unit are treated as independent (eqn (4.81)). We see that the molecule is characterized by very short runs of T, G^+, and G^-, there being only a slight change in the properties with temperature (the *gauche* states becoming more probable since x becomes closer to unity as the temperature is increased — see eqn (1.4), p. 2). The effect of including nearest-neighbour steric hindrance (eqn (4.83)) is noticeable but not particular dramatic.

TABLE 4.1.

Probability of trans *states and sequences of* trans *states for polyethylene models; the cooperative model forbids G^-G^+ and G^+G^- states*

	Probability of *trans* conformation	
j	Independent	Cooperative
270	0.555	0.659
320	0.520	0.635
370	0.494	0.617

	Probability of a sequence of j *trans* states at 323 K	
j	Independent	Cooperative
1	0.480	0.421
2	0.250	0.244
3	0.130	0.141
4	0.068	0.082
5	0.035	0.047

4.5. α-helix in polypeptides

Fig. 4.1 gives diagramatically a typical steric map[10] which shows the area in the ψ, φ plane (the two dihedral angles of internal rotation per residue shown in Fig. 3.13, p. 219) that is sterically allowed; the unallowed region is forbidden because the various groups bump for these values of ψ and φ. The α-helix of Pauling and Corey is formed when three successive residues have their angles ψ and φ simultaneously in the area

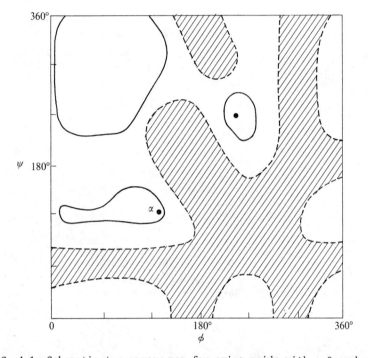

FIG. 4.1. Schematic ψ–φ energy map for amino acids with a β carbon. The shaded area represents those values of ψ and φ that lead to severe steric interaction (atom bumping) and hence is forbidden. The areas enclosed by the solid curves are the most sterically comfortable regions and are allowed. The intermediate region is sterically uncomfortable, but is not forbidden (e.g. the intermediate region is allowed if the molecule flexes somewhat by bending bond angles. The dot marked α indicates the values of ψ and φ that lead to the right-handed α-helix; the other dot indicates the values of ψ and φ that lead to the left-handed α-helix.

TECHNIQUE AND APPLICATIONS   253

of the map marked α. Fig. 4.2 illustrates the groups that
interact to form the α-helix; notice that a hydrogen bond
spans three sets of angles ψ and φ and goes from the far and
of one amide group to the far end of the other. The α-helix
can be viewed as the ordered conformation that allows the
dipoles in the amide group to line up in head-to-tail fashion
as is illustrated in Fig. 4.3.

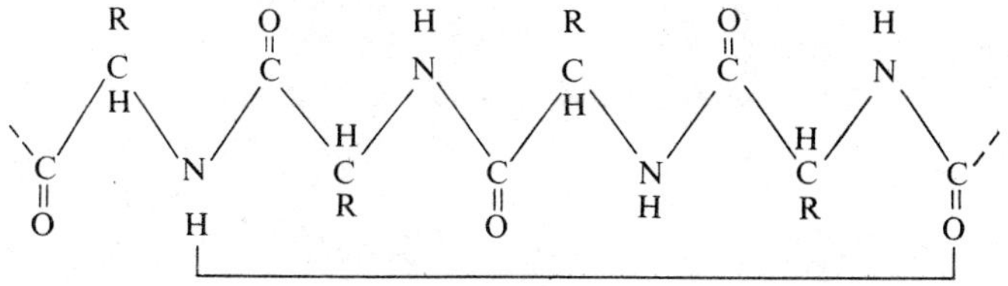

![Diagram of peptide backbone with R, C, H, N, O, H groups showing hydrogen bonding]

FIG. 4.2. Illustration of the groups that hydrogen bond in the α-helix.

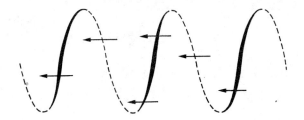

FIG. 4.3. Schematic illustration of the α-helix as a helical array of
dipoles with the dipoles lined up along the axis of the helix.

To model the equilibrium between the ordered α-helix and
all the rest of the conformations ('random coil') that do not
lead to any ordered structure (to a noticeable degree; polyethylene is referred to as a random coil yet we saw in eqn
(4.93) that it can be viewed as alternating sequences of helix,
albeit very short ones). We introduce the dichotomy that each
residue can exist in two states that are defined by φ and ψ.
The helix (h) state is designated by a small region in the
immediate neighbourhood of the dot marked α in Fig. 4.1; the
coil state (c) is all the rest of the allowed part of the ψ, φ
plane (note that the coil is thus actually a collection of

conformational states lumped together).[11] Thus we have a system where each unit (residue) can be in two states. A sample configuration of states might be

$$\ldots c\,c\,c\,c\,h\,h\,h\,c\,c\,c\,c\,h\,h\,h\,h\,h\,c\,c\,h\,h\ldots \qquad (4.94)$$

We note again that there are alternating sequences of h and c states. The simplest model we can have is one in which the free energy of a helical sequence j units long is assumed to vary linearly with j with a correction for the ends

$$g(j) = g_{ends} + j\Delta g_{unit} \quad (\text{large } j). \qquad (4.95)$$

It is convenient to take the free energy of a helix unit relative to the free energy of a coil unit; hence we write Δg_{unit}. Defining[12]

$$\sigma = \exp(-g_{ends}/RT),$$
$$s = \exp(-g_{unit}/RT), \qquad (4.96)$$

The statistical weights for coil and helix sequences are

$$u(h_j) = \sigma s^j,$$
$$u(c_j) = 1. \qquad (4.97)$$

The statistical weight for coil sequences is unity since we are taking the free energy of a unit in the helix relative to the coil.

In particular note that eqn (4.95) is a limiting expression for large j. In general an amide group in the α-helix will have a moderately strong interaction (dipole–dipole interaction) with many neighbouring amide groups, not just the one to which it is hydrogen bonded. Clearly the units at the ends have a different environment than units in the interior (having neighbouring amine groups in the nearby coil that have a more or less random orientation rather than the orderly arrangement in the α-helix). If we schematically indicate the range of the end effect by underlining the h states at the end as follows

$$\ldots c\,c\,\underline{h\,h\,h\,h\,h}\,h\,h\,h\,h\,h\,\underline{h\,h\,h\,h\,h}\,c\,c\,\ldots \qquad (4.98)$$

then for sequences longer than the combined end-effect region (in eqn (4.98) this is longer than ten residues) the end effects are the same. For sequences shorter than the range of end effects described by g_{ends}, e.g.

$$\ldots c\,c\,\underline{h\,h\,h\,h\,h\,h\,h}\,c\,c\,\ldots, \qquad (4.99)$$

the factor g_{ends} used to describe the end regions in eqn (4.98) does not apply, and we would have to assign special statistical weights to all the sequences whose lengths are shorter than the combined range of the end effects on both ends. We will ignore this complication, using eqn (4.95) or eqn (4.97) to describe helical sequences for all j. Of course we must justify this procedure and we will return shortly to this point.

Note that eqn (4.95) does not assume that the statistical weights are composed simply of nearest-neighbour (or nearest amide-group) interaction, but states that the free energy is linear in sequence length with a correction for the fact that units on or near the end have a different environment. Thus both g_{ends} and Δg_{unit} can include interactions with many neighbouring units. Assuming that eqn (4.97) is valid for all values of j reduces the mathematical problem to the exact analogue of a nearest-neighbour model, but we must remember that, as far as the interpretation of the statistical weights is concerned, the model is pseudo-nearest-neighbour in that the possibility of interaction over many neighbours is included.[13]

We examine first the properties of long chains. Using eqn (4.97) in eqn (4.50) we have

$$U_c(\lambda) = \sum_{j=1}^{\infty} (1/\lambda)^j = \frac{1}{\lambda - 1}$$

$$U_h(\lambda) = \sigma \sum_{j=1}^{\infty} (s/\lambda)^j = \frac{\sigma s}{\lambda - s} \qquad (4.100)$$

(cont.)

$$\frac{\sigma s}{(\lambda-1)(\lambda-s)} = 1$$

or

$$\lambda_1 = \tfrac{1}{2}[(1+s) + \{(1-s)^2 + 4\sigma s\}^{\tfrac{1}{2}}]. \tag{4.101}$$

The fractions of helix, coil, and states that initiate helical sequences (a factor σ is assigned to every helical sequence, hence the logarithmic derivative of λ_1 with respect to σ gives the fraction of states that initiate helical sequences) are

$$\vartheta_h = 1 - \vartheta_c = \tfrac{1}{2}\left[1 + \frac{(s-1)}{\{(s-1)^2 + 4\sigma s\}^{\tfrac{1}{2}}}\right], \tag{4.102}$$

$$\vartheta_{seq} = 4\sigma s / \left[1+s+\{(s-1)+4\sigma s\}^{\tfrac{1}{2}}\right]\{(s-1)+4\sigma s\}^{\tfrac{1}{2}}.$$

The average-length helical sequence is given by

$$\langle L_h \rangle = \vartheta_h / \vartheta_{seq}. \tag{4.103}$$

At $s = 1$ (where $\Delta g_{unit} = 0$, or the free energy of a unit in the helical state is equal to that of a unit in the coil state) we have

$$\vartheta_h = \tfrac{1}{2} \text{ (independent of } \sigma\text{),}$$
$$\langle L_h \rangle = 1 + \sigma^{-\tfrac{1}{2}}. \tag{4.104}$$

We also have the useful relation (ignoring the temperature dependence of σ)

$$\left(\frac{\partial \ln \vartheta}{\partial (1/T)}\right)_{\vartheta=\tfrac{1}{2}} = -\frac{\Delta h_u \sigma^{-\tfrac{1}{2}}}{2R}, \tag{4.105}$$

where

$$s = \exp(-\Delta h_u/RT)\exp(\Delta s_u/R) = \exp(\Delta h_u(T-T_0)/RTT_0)$$

$$\frac{\partial \ln s}{\partial (1/T)} = -\frac{\Delta h_u}{R},$$

(4.106)

where the second equality for s holds if Δh_u and Δs_u are taken as independent of temperature.

We expect the following magnitudes for the parameters s and σ

$$\left.\begin{array}{l}\text{both helix and coil}\\ \text{in equilibrium}\end{array}\right\} \quad g_{unit} \sim 0, \quad s \sim 1;$$

$$\left.\begin{array}{l}\text{ends of a helical}\\ \text{sequence higher in}\\ \text{free energy than}\\ \text{interior units; ends}\\ \text{represent defects}\end{array}\right\} \quad g_{ends} = \text{(positive)}, \quad \sigma \ll 1.$$

(4.107)

From eqn (4.108) we see that $\vartheta = \frac{1}{2}$ at $s = 1$, independent of the value of σ. Thus the temperature where $\vartheta = \frac{1}{2}$ is T_0. If we measure the derivative in eqn (4.105) experimentally for long polypeptide chains, then knowing T_0, we have one equation in two unknowns, Δh_u and σ. We will see shortly that the measurement of $\vartheta(T=T_0)$ for a short chain is sufficient to give σ, thus giving Δh_u from eqn (4.105). Alternatively we could measure Δh_u calorimetrically. In Chapter 6 we will discuss all the parameters Δh_u, T_0 and σ that are known. Here I quote typical orders of magnitude to use in our discussion

$$T_0 \sim 303 \text{ K},$$
$$\Delta h_u \sim 1.25 \text{ kJ mol}^{-1}, \quad (4.108)$$
$$\sigma \sim 10^{-4}.$$

Thus the transition between helix and coil occurs in an observable range (between 273 K and 373 K for aqueous solution). Note that Δh_u is very small, representing the fact that Δh_u measures the difference in enthalpy between an amide hydrogen bond and the water–amide hydrogen bond (see section 2.2.12). The parameter σ is seen to be quite small, a fact that will be seen to be all important in determining the nature of the equilibrium; for σ to be 10^{-4} requires $g_{ends} \sim 23$ kJ mol^{-1}, which is not excessively large. We have ignored the temperature dependence of σ in the above discussion; this could be included by the results are very insensitive to small variations in σ.

We now want to explore the nature of the equilibrium between helix and coil in large chains using the parameters of eqn (4.108). At $s = 1$ half the units are in the helix state and half are in the coil state. The average length helical sequence is given by eqn (4.104); we have

$$(\sigma = 1) \quad \langle L \rangle = 2,$$
$$(\sigma = 10^{-4}) \quad \langle L \rangle \simeq 100. \tag{4.109}$$

The case $\sigma = 1$ applies approximately to the stacking equilibrium in single-strand polynucleotides discussed in section 4.1. At $\vartheta = \frac{1}{2}$ in that system the average run of stacked units is only two (alternating on the average with two unstacked units). A typical configuration of states for a system at $s = 1$ with $\sigma = 1$ is given by the sequence of heads and tails obtained from throwing pennies. We see that for $\sigma = 10^{-4}$ the average helix sequence is very long. To understand why the sequences tend to be long consider the equilibrium at $s = 1$

$$\ldots c\ c\ h\ h\ h\ h\ c\ c\ c\ h\ h\ h\ h\ h\ c\ c\ c\ \ldots$$
$$\ldots c\ c\ c\ c\ h\ h\ h\ h\ h\ h\ h\ h\ h\ c\ c\ c\ c\ c\ \ldots \tag{4.110}$$

On the left-hand side of the equilibrium there are two short sequences each contributing a factor σ; the net statistical weight of the configuration is thus σ^2. On the right-hand side there is only one helical sequence and one factor σ. Since $\sigma \ll 1$, short sequences will tend to coalesce into one large sequence, thus reducing the number of σ factors (or reducing the number of sequence ends which are unfavourable). This is termed a cooperative equilibrium since if the sequences 'cooperate' and merge, thus sharing a common σ factor, the number of unfavourable helix ends is reduced.

The equilibrium in eqn (4.110) can be expressed quantitatively in terms of the combinatorial expression of eqn (4.76) which in the model of eqn (4.97) becomes

$$q = \sum_{N_h} \sum_{N_s} \frac{N_h!}{(N_h-N_s)!N_s!} \frac{(N-N_h)!}{(N-N_h-N_s)!N_s!} s^{N_h} \sigma^{N_s}, \quad (4.111)$$

N_h being the number of h states and N_s the number of helical sequences. At $s = 1$ the only factors in eqn (4.111) are the combinatorial factor and σ^{N_s}. Considered alone, the combinatorial factor has its maximum value when $N_h = N/2$ and $N_s = N/4$ (the most probable values for independent units). Thus the combinatorial factor favours many short sequences. Since $\sigma \ll 1$, σ^{N_s} becomes smaller the larger N_s (the more helical sequences there are). The quantity σ^{N_s} is a maximum when N_s is a minimum (zero). Thus there is a balance between many short sequences which maximizes the combinatorial factor (combinatorial entropy) and few long sequences which makes σ^{N_s} larger (less small). The balance of this equilibrium is given by eqn (4.104), $\langle L \rangle = 1 + \sigma^{-\frac{1}{2}}$.

We have seen that when σ is very small the average helical sequence length is large. The distributions of helical lengths are given by eqn (4.68) and (4.70)

$$p(h_j) = (s/\lambda_1)^j / \sum_{j=1}^{\infty} (s/\lambda_1)^j = \frac{(\lambda_1-s)}{s}\left(\frac{s}{\lambda_1}\right)^j, \quad (4.112)$$

$$P(h_j) = j(s/\lambda_1)^j / \sum_{j=1}^{\infty} j(s/\lambda_1)^j = (1-s/\lambda_1)^2 j\left(\frac{s}{\lambda_1}\right)^j.$$

As with the distribution of eqns (4.91) and (4.92), (s/λ_1) is necessarily less than unity. Thus $p(h_j)$ (the probability a helical sequence picked at random has j units) is a monotonically decreasing function of j (the most probable sequence length being one). At $s = 1$ eqn (4.101) gives

$$\lambda_1 = 1 + \sigma^{\frac{1}{2}} \quad (4.113)$$

Thus $s = 1$ the quantity $s/\lambda_1 = 1/(1+\sigma^{\frac{1}{2}})$ which for $\sigma = 10^{-4}$ is $(1/1.01)$; hence $(1/1.01)^j$ drops off with j very slowly. The function $P(h_j)$, the probability that a helical unit picked at random will be in a helical sequence of j units, has a maximum determined by

$$\frac{\partial P(h_j)}{\partial j} = 0, \qquad (4.114)$$

giving (see eqn (1.21))

$$j_{max} = \frac{1}{\ln(\lambda_1/s)}, \qquad (4.115)$$

$$(\text{at } s = 1) \quad j_{max} \simeq \sigma^{-\frac{1}{2}} \simeq \langle L_h \rangle.$$

A measure of the width of the distribution $P(h_j)$ is given by $\{\langle j^2 \rangle - \langle j \rangle^2\}^{\frac{1}{2}}$. We have (for N infinite)

$$\langle j \rangle = \lambda_1/(\lambda_1 - s),$$

$$\langle j^2 \rangle = \lambda_1(\lambda_1 + s)/(\lambda_1 - s)^2, \qquad (4.116)$$

$$\Delta = (\langle j^2 \rangle - \langle j \rangle^2)^{\frac{1}{2}} = \sqrt{(\lambda_1 s)}/(\lambda_1 - s),$$

$$(\text{at } s = 1) \quad \Delta \simeq \sigma^{-\frac{1}{2}} \simeq \langle L_h \rangle.$$

Thus $P(h_j)$ at $s = 1$ is a very broad distribution with a maximum at $\langle L_h \rangle$ and a distribution width also of $\langle L_h \rangle$.

Fig. 4.4 shows $p(h_j)$ and $P(h_j)$ (at $s = 1$ these are the same as $p(c_j)$ and $P(c_j)$) calculated for $\sigma = 10^{-4}$. Using Fig. 4.4 we return to the question of the validity of using eqn (4.97) for all j. The fraction of h states in helical sequences shorter than 10 units is shown by the shaded region in Fig. 4.4. This fraction is seen to be very small. Hence, even though eqn (4.97) treats short sequences incorrectly, the number of h states in these sequences is completely negligible for $\sigma \ll 1$. Thus for very cooperative models (small σ), the use of eqn (4.97) for all j is justified.

Since we have seen that the average length of a helical sequence at $s = 1$ is of the order of 100 units, it is clear that when the chain length is of the order of only several hundred units the finiteness of the chain will start to

TECHNIQUE AND APPLICATIONS 261

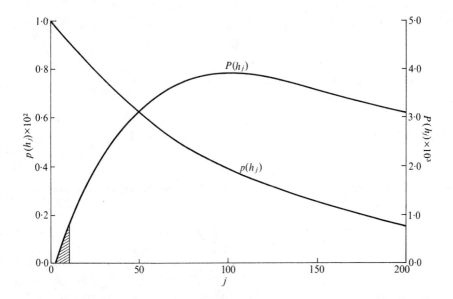

FIG. 4.4. The probability $p(h_j)$ that a given helical sequence contains j units, and the probability $P(h_j)$ that a given helix unit is in a sequence of j units for an infinite polypeptide with $s = 1$ and $\sigma = 10^{-4}$. The shaded area gives the fraction of units in sequences with $j < 10$.

influence the equilibrium. To treat finite chains we turn to the matrix method. The table of (4.11) using the model of eqn (4.97) is

$$
\begin{array}{c|cc}
 & \multicolumn{2}{c}{i+1} \\
i & h & c \\
\hline
h & s & \sigma s \\
c & 1 & 1
\end{array}
\qquad (4.117)
$$

giving the partition function as

$$q = [1 \ 1] \begin{bmatrix} s & \sigma s \\ 1 & 1 \end{bmatrix}^N \begin{bmatrix} 0 \\ 1 \end{bmatrix} \qquad (4.118)$$

where the end vectors are constructed as for eqn (4.86) (last unit on the 'right end' treated as if it is always followed by a c state). The eigenvalues are given by

$$\begin{vmatrix} s-\lambda & \sigma s \\ 1 & 1-\lambda \end{vmatrix} = 0, \qquad (4.119)$$

which gives eqn (4.100). The partition function written in terms of the eigenvalues is

$$q = c_1 \lambda_1^N + c_2 \lambda_2^N, \qquad (4.120)$$

where λ_1 and λ_2 are the solutions of eqn (4.119). The c factors can be easily constructed using eqn (4.62). With U_h of eqn (4.100) we have

$$c_i(\lambda_i) = \lambda_i / \left\{ 1 + \frac{\sigma s}{(\lambda_i - s)^2} \right\}. \qquad (4.121)$$

The average properties, now a function of N, are calculated in the usual manner, e.g.

$$\vartheta_h = \frac{1}{N} \frac{\partial \ln q}{\partial \ln s}. \qquad (4.122)$$

Since the $\lambda_i(s,\sigma)$ and the $c_i(s,\sigma)$ are given as explicit functions of s and σ all the derivatives can be calculated, the procedure being straightforward, albeit tedius. I quote here the results; the details are given elsewhere.[14] We define a scaled chain length η and a scaled temperature τ as

$$\eta = \sigma^{\frac{1}{2}} N,$$
$$\tau = (T_0 - T) \Delta h_u \sigma^{-\frac{1}{2}} / 4RT_0^2. \qquad (4.123)$$

For σ small at $s = 1$, $\langle L \rangle = 1 \div \sigma^{-\frac{1}{2}} \simeq \sigma^{-\frac{1}{2}}$; hence eqn (4.123) can be written

$$\eta = N / \langle L \rangle$$
$$\tau = (T_0 - T) \langle L \rangle \Delta h_u / 4RT_0^2. \qquad (4.124)$$

In terms of the parameters η and τ we obtain for small σ at

$s = 1$,

$$\vartheta_h(\eta) = \tfrac{1}{2}(1 - \eta^{-1} \tanh \eta) = 1 - \vartheta_c(\eta), \qquad (4.125a)$$

$$\frac{\partial \vartheta_c(\eta)}{\partial \tau} = 2\vartheta_h(\eta) \tanh \eta. \qquad (4.125b)$$

In addition, we can derive an explicit relation for the fraction of coil units at the ends of the molecule

$$\vartheta_{c\text{-end}} = \frac{(2/\eta) \sinh \eta - 1}{\cosh \eta}. \qquad (4.126)$$

Using eqn (4.125b) to give the total fraction of coil (coil sequences at the ends plus interior coil sequences) we have the ratio

$$\frac{\vartheta_{c\text{-end}}}{\vartheta_{c\text{-total}}} = 2\left(\frac{2 \sinh \eta - \eta}{\eta \cosh \eta + \sinh \eta}\right). \qquad (4.127)$$

We find that the value of the chain length $(2N\sigma^{\frac{1}{2}})$ corresponding to $\eta = 2$ is a characteristic value. We have

$$\vartheta_h(\eta) \simeq \frac{\eta^2/6}{\eta^2/2 + 1} \qquad (\eta < 2),$$

$$\vartheta_h(\eta) \simeq \tfrac{1}{2}(1-\eta^{-1}) \qquad (\eta > 2), \qquad (4.128)$$

$$\frac{\vartheta_{c\text{-end}}}{\vartheta_{c\text{-total}}} \simeq 1 \qquad (\eta < 2).$$

The first of eqn (4.128) arises from truncation of the expansion for tan hη. The last of eqn (4.128) indicates that for $\eta < 2$ (or $N < 2 \langle L \rangle$) all the unwinding is from the ends of the molecule. This is reasonable since, if $\langle L \rangle \sim 100$, for a chain length of 200 or less there will on the average only be one long helical sequence (hence all unwinding is from the ends). The partition function for the model where only one helical sequence of any length is allowed on a chain of N units follows the development given in section 3.1. We have

$$q = 1 + \sigma \sum_{n=1}^{N} (N-n+1)s^n. \qquad (4.129)$$

The difference between the system discussed in section 3.1, and the one-sequence model for polypeptides is that the all-coil state (the '1' in eqn (4.129)) does not involve the dissociation of strands. Applying eqn (4.122) to the partition function of eqn (4.129) we readily find at $s = 1$ for small σ

$$\vartheta_h(\eta) = \frac{\eta^2/6}{\eta^2/2 + 1}, \qquad (4.130)$$

which is identical to the first eqn (4.128) derived from truncating the expansion of tanh η.

Fig. 4.5 shows $\vartheta_h(\eta,\tau)$. Note that in terms of η and τ all

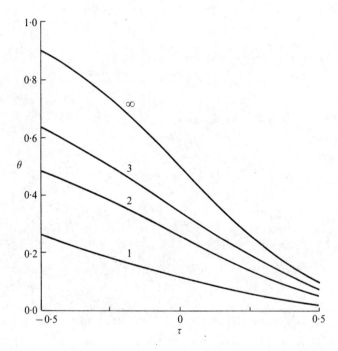

FIG. 4.5. The fraction of helix in polypeptides as a function of the parameters $\eta = 1, 2, 3$, and ∞ and τ. Note that in terms of η and τ all polypeptides have the behaviour shown in this figure.

systems following the model of eqn (4.97) for small σ have exactly the same behaviour. We see that helix is less stable (melts out at lower temperature) in short chains. This is simply because helical sequences of length $\langle L \rangle$ have less combinatorial entropy on short chains or, in other words, finiteness of the chains cramps the formation of long sequences. From eqn (4.105) for long chains

$$\left(\frac{\partial \ln \vartheta_h}{\partial 1/T}\right)_{\vartheta = \frac{1}{2}} = -\tfrac{1}{2}\langle L \rangle \Delta h_u. \qquad (4.131)$$

Thus as the model becomes more cooperative (σ small, $\langle L \rangle$ large) the transition becomes sharper. From eqn (4.131) we see that the temperature variation of ϑ at $\vartheta = \tfrac{1}{2}$ is given by the enthalpy change of the average structure at $s = 1$; as σ becomes smaller, the average sequence becomes larger and the transition sharper.

There are three parameters required to treat the helix–coil transition in polypeptides: T_0, Δh_u, and σ. T_0 is obtained as the temperature where $\vartheta_h = \tfrac{1}{2}$ for long chains. From eqn (4.15a) (or to a very good approximation of one eqn (4.128)), a single measurement of ϑ_h for a chain length (preferably in the neighbourhood of $\eta = 2$) at $s = 1$ yields the value of σ. Then from measurement of the variation of ϑ with temperature for the long chain, eqn (4.105) gives Δh_u. We find that 'very long' or 'infinite' chains practically means $\eta > 10$ (if $\sigma = 10^{-4}$ this means $N > 1000$).

We close our remarks on α-helix in polypeptides by noting again that the model is given by eqn (4.95) which assumes that for large sequence lengths the free energy of a sequence is linear in sequence length with a correction factor for the ends. We need not assume that the interactions in the helix or in the coil are limited to near neighbours only. The justification of the use of eqn (4.95) (or eqn (4.97)) for all sequence lengths is that in a highly cooperative process very few units are in short sequences, hence treating such sequences admittedly incorrectly makes no difference that is observable experimentally.

4.6. Long-chain synthetic polynucleotides

In Chapter 3 we treated the association equilibrium of short synthetic polynucleotides and polypeptides and showed that useful information (thermodynamic parameters) could be obtained from such systems. In the present section we explore the other limit, that of infinite polynucleotides. While for short chains all unwinding of the double helix was by unravelling from the ends (and by complete dissociation of the complex), for infinite chains all unwinding of the double helix must be by the formation of interior loops (since the chains are infinite the ends cannot play a role in the net fraction of unwinding). This limit, unlike the limit of short chains, is not particularly useful since the limit of infinite chains is unrealistic. While a chain length of 5000 units or so may seem long, as we saw in the last section, a more relevant parameter is the ratio of the actual chain length to the correlation length (average length of a helical sequence); for very cooperative transitions the correlation length can be quite long (~ 100 units in polypeptides). As we have already discussed, the entropy of loops is lower than that of free chains such as are formed by unwinding from the ends. Thus free ends have a bias over loops. For the above reasons it turns out that the relations we will give in this section are not useful in the interpretation of real systems. None the less they are useful in giving the limiting case of no unwinding from the ends, and illustrate the nature of the transition when it takes place by a loops only mechanism. In real DNA, as we might guess, the nature of the transition is about halfway between the behaviour of short chains (all unwinding from the ends) and infinite chains (all unwinding by internal loops); we will discuss realistic models for DNA in Chapter 6. In addition, the model of an infinite polynucleotide has a fascinating property: it exhibits a true phase transition. While a true phase transition does not occur in real (finite) chains, the study of how a true phase transition arises in infinite chains illuminates the nature of the cooperative transition.

Fig. 4.6 illustrates the two models we will discuss. The first is the perfect-matching model, where the two chains must

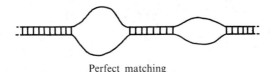

Perfect matching

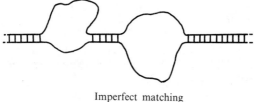

Imperfect matching

FIG. 4.6. Illustration of perfect and imperfect matching in double-strand polynucleotides.

be in register in helical regions with the result that when a loop is formed both chains in the loop have the same length. Of course, the cause of the chains being in perfect register is the fact that they have complimentary specific sequences, and the requirement of Watson–Crick pairing allows them to form extensive double helix in only one way. A result of specific sequence is that the stability of the helix is not homogeneous (A–T pairs being weaker helix formers than G–C pairs). We will treat this effect explicitly in Chapter 6; here we use the perfect-matching case assuming homogeneous helix stability (i.e. that we can assign an average s factor to each unit) solely as an interesting model that sheds light on the effect of loops on the transition. In synthetic polynucleotides that form double helix each unit in the chain is the same and there is thus nothing to keep the chains in perfect register, allowing loops to form that do not have equivalent contributions from each strand (imperfect-matching model). We turn first to the perfect-matching model.

We have already discussed the statistical weights for a system having loops. With eqn (3.88) of section 3.2.2 the analogue of eqn (4.97) (polypeptide model) is

$$u(h_j) = \sigma s^j, \qquad (4.132)$$
$$u(c_j) = 1/j^c.$$

Using the statistical weights of eqn (4.132) in eqn (4.50) yields

$$\frac{\sigma s}{\lambda - s} \sum_{j=1}^{\infty} j^{-c}\left(\frac{1}{\lambda}\right)^j = 1. \qquad (4.133)$$

While the sum in eqn (4.133) cannot be evaluated exactly, Fisher[16] has shown that at $\lambda \to 1$ a very good approximation is given by

$$\sum_{j=1}^{\infty} j^{-c}\left(\frac{1}{\lambda}\right)^j \simeq \left(\sum_{j=1}^{\infty} j^{-c}\right)\left\{1 - \left(\frac{\lambda-1}{\lambda}\right)^{c-1}\right\}. \qquad (4.134)$$

From the polypeptide model we have seen (eqn (4.113)) that $\lambda_1 = 1 + \sigma^{\frac{1}{2}}$ at $s = 1$, indicating that for highly cooperative systems λ_1 is close to unity in the transition region. Since λ_1 and s are both expected to be close to unity in the transition region it is convenient to define

$$\lambda = 1 + \varepsilon,$$
$$\qquad (4.135)$$
$$s = 1 + \delta,$$

and work with the small quantities ε and δ instead of λ and s. Defining

$$\sigma' = \sigma \sum_{j=1}^{\infty} j^{-c}, \qquad (4.136)$$

eqn (4.133) (using eqns (4.134), (4.135), and (4.136)) becomes

$$\text{(perfect matching)} \qquad \sigma'(1-\varepsilon^{c-1}) = \varepsilon - \delta, \qquad (4.137a)$$

$$\text{(polypeptide)} \qquad \sigma\varepsilon^{-1} = \varepsilon - \delta, \qquad (4.137b)$$

where we have used the definitions of eqn (4.135) to give the corresponding relation for polypeptides (eqn (4.100)), which is the special case of eqn (4.133) with $c = 0$). In eqn (4.137) we have taken terms like $(1 + \varepsilon)$ and $(1 + \delta)$ as unity since the dominating factors are ε and δ appearing as factors.

In eqn (4.137b) it is seen that $\varepsilon = 0$ is a possible solution of eqn (4.137a). We have

$$(\varepsilon = 0) \qquad \delta = -\sigma'. \qquad (4.138)$$

Thus as $s = 1 - \sigma'$, $\varepsilon = 0$, and $\lambda_1 = 1$. This means that at this temperature the partition function per unit is identically equal to the statistical weight of a free coil unit (we take the free coil as the reference, hence the statistical weight of the coil is unity). A solution $\lambda_1 = 1$ means that we have a pure coil phase at a finite temperature. To have 100 per cent of one state at a finite temperature (no mixing of the other states allowed) is qualitatively different from the transition in polypeptides where the helix–coil transition is sharp, but where the limits of all helix or all coil are approached asymptotically for temperatures above or below T_0 respectively. To explore the nature of the transition for the perfect-matching model in the neighbourhood of $\varepsilon = 0$, we note that for ε and δ small

$$\vartheta_h = \frac{\partial \ln \lambda_1}{\partial \ln s} \sim \frac{\varepsilon}{\delta} \sim \varepsilon^{2-c}. \qquad (4.139)$$

Taking T_c as the temperature where $\varepsilon = 0$ (which from eqn (4.138) is not the temperature T_0, where $s = 1$ or $\delta = 0$) and expanding ε in a series in $(T-T_c)$ about T_c gives[16]

$$\vartheta_h \sim (T-T_c)^{(2-c/c-1)}. \qquad (4.140)$$

The result of eqn (4.140) is that for $c > 1$, ϑ_h goes exactly to zero at T_c; for $c > \frac{3}{2}$ ϑ_h goes to zero at T_c with an infinite slope. From eqn (3.90) we have for real chains that $c \simeq 1.75$. Thus the perfect-matching model for infinite chains shows a phase transition. The nature of the transition for the perfect-matching model is shown in Fig. 4.7 (solid curve). The reason for the transition is that the unfavourable loop entropy, reflected in the factor j^{-c} in eqn (4.133), makes the

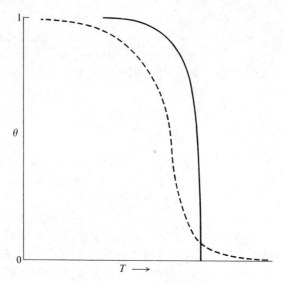

FIG. 4.7. ϑ as a function of temperature for the perfect-matching model (solid curve) and the imperfect-matching model (dashed curve).

mixing of coil and helix (via loop formation) of such unfavourable free energy that the system goes to the all coil state.

In section 1.9 we discussed spherical physical clusters. In that model a phase transition also appeared. We compare the two cases below,

$$\text{(sum at } x = 1\text{)} \quad \Sigma y^{n^{2/2}} \quad \text{(physical clusters)},$$
$$\text{(sum at } \lambda = 1\text{)} \quad \Sigma j^{-c} \quad \text{(perfect matching)}. \tag{4.141}$$

In the physical-cluster model a phase transition occurred because the sum in (4.141) converged at $x = 1$; the reason for the convergence was the fact that if the surface free energy was unfavourable the factor y was less than unity. In the perfect-matching model the sum over j^{-c} converges at $\lambda = 1$ for $c > 1$, the form j^{-c} coming from a long-range correction to the entropy of chains in loops.

The imperfect-matching model introduces one new feature. Since the number of units from each strand in a loop do not need to be equal in the imperfect-matching model, there are

j ways a total of j units can be distributed between the two halves of the loop. Thus in the imperfect-matching model the generating function for coil states becomes

$$U_c(\lambda) = \sum_{j=1}^{\infty} \frac{j}{j^c}\left(\frac{1}{\lambda}\right)^j = \sum_{j=1}^{\infty} \frac{1}{j^{c-1}}\left(\frac{1}{\lambda}\right)^j. \tag{4.142}$$

For $c = 1.75$ the sum $\Sigma 1/j^{c-1}$ (eqn (4.142) with $\lambda = 1$) diverges and hence there is no phase transition (i.e. $\lambda_1 = 1$ corresponding to the pure coil state is no longer an allowed solution). Physically the reason the phase transition is lost in the imperfect-matching model is that the combinatorial entropy of loop function (number of ways loops with unequal sides can be formed) makes up some of the entropy lost in loop formation. The dashed curve in Fig. 4.7 shows the form of the transition in the imperfect-matching model, indicating that ϑ_h goes to zero asymptotically as the temperature goes to infinity, unlike the perfect-matching model case. For both the perfect-matching model and the imperfect-matching model the transitions are noticeably asymmetric unlike the transition curve for long polypeptides (Fig. 4.5).

4.7. Evaluation of derivatives of partition functions [17, 18]

For a system with arbitrary interactions with a finite number of neighbouring units, the partition function can be written as a matrix product. For this general class of models the partition function is given in terms of the eigenvalues of the appropriate matrix by eqn (4.38) with average quantities given by the appropriate analogue of eqn (4.122). For matrices larger than 2 × 2 this procedure is difficult to follow since we cannot obtain an explicit solution for the eigenvalues. If the eigenvalues are determined numerically the appropriate derivative of the eigenvalues could be determined numerically (by calculating the eigenvalues at close intervals of the appropriate statistical weight and taking finite differences). This is not necessary if we have the secular equation in the form (e.g. see eqn (4.87) or eqn (4.100))

$$\sum_{n=0}^{\infty} A_n(s)\lambda^n = 0, \qquad (4.143)$$

where the $A_n(s)$ are factors composed of the statistical weights for the model, illustrated here by the parameters; ρ is the size of the matrix required for the model. Then the derivative of λ with respect to s is given by implicit differentiation of eqn (4.143)

$$\left(\frac{\partial \ln \lambda}{\partial \ln s}\right)_{\lambda=\lambda_i} = \frac{-\sum_{n=0}^{\rho} \frac{\partial A_n(s)}{\partial s}\lambda_i^n}{\sum_{n=1}^{\rho} nA_n(s)\lambda_i^n} . \qquad (4.144)$$

Eqn (4.62) also given an explicit form for the $c_i(\lambda_i)$ factors. Using eqn (4.144) and eqn (4.62) the only numerical problem is to determine the roots of eqn (4.143). This is often complicated by the fact that for many models the roots are complex (the largest root always being real). We now show that any order derivative can be calculated as a simple matrix product.

Taking s as a general parameter which in the model is assigned in a one-to-one fashion with the states of interest (if there is no such parameter in the model we can introduce one and set it equal to unity after differentiating with respect to it). Then the first derivative of the partition function expressed as a matrix product is

$$\frac{\partial q}{\partial \ln s} = \mathbf{e}\left(\sum_{M=0}^{N-1} \mathbf{W}^{N-M-1}\mathbf{W}'\mathbf{W}^M\right)\mathbf{e}^+, \qquad (4.145)$$

where

$$\mathbf{W}' = \partial \mathbf{W}/\partial \ln s. \qquad (4.146)$$

Defining the hypermatrix H,

$$\mathbf{H} = \begin{bmatrix} \mathbf{W} & \mathbf{W}' \\ \mathbf{0} & \mathbf{W} \end{bmatrix} \qquad (4.147)$$

we observe that

$$\frac{\partial q}{\partial \ln s} = e(h_{1,2})e^+, \quad (4.148)$$

where $h_{1,2}$ is the (1,2) element of the hypermatrix H. Eqn (4.147) can be generalized to

$$H = \begin{bmatrix} W & W' & 0 & 0 & 0 \\ 0 & W & W' & 0 & 0 \\ 0 & 0 & W & W' & 0 \\ 0 & 0 & 0 & W & W' \\ 0 & 0 & 0 & 0 & W \end{bmatrix} \quad (4.149)$$

etc., giving

$$\frac{\partial^n q}{\partial (\ln s)^n} = e(h_{1,n+1})e^+. \quad (4.150)$$

Eqn (4.149) makes use of the fact that for linear chain models the matrix gives the states of one unit in terms of the states of neighbouring units giving the property

$$\frac{\partial^n W}{\partial (\ln s)^n} = \frac{\partial W}{\partial \ln s}. \quad (4.151)$$

Further details of this procedure and generalizations are given elsewhere.[18] The utility of this approach is that any order derivative of the partition function for a finite chain can be calculated as a simple matrix product, an extremely simple task for a computer. Thus the numerical calculation of eigenvalues is completely avoided.

Notes and References

1. A more detailed discussion of some of the points in this chapter is found in P&S, with particular reference to comparisons of formulations by different authors. In P&S a great deal more detail is given for the matrix method (illustrating the construction of matrices, etc.). Section 4.2 is new, however, and permits a shorter and neater derivation for many results than is given in P&S.
2. Poland, D., Vournakis, J.N., and Scheraga, H.A., *Biopolymers* **4**, 223 (1966).
3. Not all matrices are diagonizable. I have, however, never encountered an example, representing a physical model, that was not diagonizable.

4. P&S, Appendix A.
5. Poland, D., *J. chem. Phys.* **60**, 808 (1974).
6. The application of generating functions to biopolymers was originally presented by Lifson in a different manner: Lifson, S., *J. chem. Phys.* **40**, 3705 (1964).
7. Peller has given a similar partition function: Peller, L., *J. phys. Chem.* **63**, 1194 (1959).
8. Hill, T.L., *J. chem. Phys.* **30**, 383 (1959).
9. For a discussion of the rotational isomers in n-alkanes and many references to the statistical mechanics of polyethylene see: Scott, R.A. and Scheraga, H.A., *J. chem. Phys.* **44**, 3054 (1966). For a discussion of cyclic alkanes see the second entry in ref. 19, Chapter 2.
10. The map shown is an approximate composite representing the features on which most authors agree. For references see ref. 18, Chapter 2.
11. The definitions we use of h and c states are due to Lifson and Roig: Lifson, S. and Roig, A., *J. chem. Phys.* **34**, 1963 (1961).
12. The parameters σ and s are often referred to as the Zimm–Bragg parameters after the authors that first introduced them: Zimm, B.H. and Bragg, J.K., *J. chem. Phys.* **31**, 526 (1959).
13. In most of the literature on the helix–coil transition, e.g. refs. 11 and 12, we find a very specific interpretation of σ and s (or analogous parameters) in terms of hydrogen bonds, residue entropy, etc. I must emphasize that this specific interpretation is overly restrictive.
14. See P&S, p. 94.
15. Ref. 3, Chapter 3.
16. Ref. 7, Chapter 3.
17. Application of this technique are illustrated in: Poland, D. and Scheraga, H.A., *Biopolymers* **7**, 887 (1969).
18. Springgate, M. and Poland, D., *J. chem. Phys.* **62**, 675 (1975).

5

COOPERATIVE BINDING TO BIOPOLYMERS

The study of the binding of small molecules to biological macromolecules is a common procedure used to gain information about the number and nature of acceptor sites on a macromolecule. In this chapter I shall show how cooperative binding can be treated in a systematic manner using several explicit examples.

5.1. Titration of rigid macromolecules

We once again return to the model of a polyprotic acid containing N sites for the binding of protons. This model has already been discussed in section 1.7 for the case of independent sites and in section 4.1 for the case of nearest-neighbour repulsive interactions between charged sites. For simplicity we will focus attention on a rigid linear macromolecule of regular structure; at the end of this section we will indicate how to treat titrations when the molecule has the option of changing conformation to more comfortably accommodate the charge density (charge induced helix—coil transition).

An example of such a system is the titration of polylysine in a 95 per cent methanol/5 per cent water mixture.[1] In this solvent polylysine exists in the α-helix conformation and remains so as it is completely titrated. The groups that dissociate protons are the ε-amino groups of the lysine sidechains. A model compound for the titration is n-butylamine

$$CH_3(CH_2)_3NH_3^+ \rightleftharpoons CH_3(CH_2)NH_2 + H^+ \quad (pK = 10.1), \quad (5.1)$$

where the pK value quoted is for 95 per cent methanol (which is seen to be similar to the pK for amino groups in water given in Table 1.3). Fig. 1.6 illustrates the nature of the equilibrium between charged and uncharged forms for methylamine in water. Fig. 5.1 shows the experimental titration curves for n-butylamine and polylysine (degree of polymerization approximately 1600 residues). The experimental titration curve for

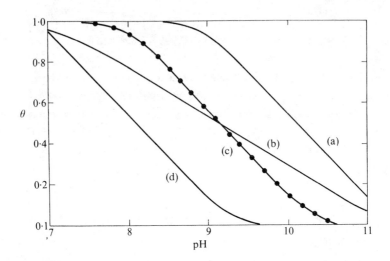

FIG. 5.1. Titration curves in 95 per cent MeOH: (a) n-butylamine; (b) polylysine. The other curves are explained in the text.

n-butylamine is given very accurately by eqn (1.68). The experimental titration curve for polylysine is seen to be shifted to lower pH values and to be broader than that for n-butylamine. We explain this shift by simple electrostatic repulsion. Thus for the reaction

$$\underbrace{NH_3^+ \quad NH_3^+}_{} \rightleftharpoons \underbrace{NH_2 \quad NH_3^+}_{} + H^+ \qquad (5.2)$$

the loss of a proton by an $-NH_3^+$ group is made easier by the fact that such a loss relieves the electrostatic repulsion of neighbouring $-NH_3^+$ groups when they exist side by side on a rigid macromolecule. Looking at the reaction as going from right to left, it is harder to bind a proton for a group neighbouring a charged group (proton already bound), thus requiring a higher hydrogen ion concentration (lower pH) to make the reaction take place. We now give a quantitative model for this effect.

First we review some facts about proton binding that we have already discussed. For independent units

COOPERATIVE BINDING TO BIOPOLYMERS

$$\begin{pmatrix} \text{Statistical weight} \\ \text{of unbound group} \end{pmatrix} = 1,$$

$$\begin{pmatrix} \text{Statistical weight} \\ \text{of bound group} \end{pmatrix} = [H^+]K = 10^{pK-pH} = x.$$

(5.3)

A general relation for the fraction (ϑ) of bound protons is

$$\vartheta = \frac{1}{N}\frac{\partial \ln \xi}{\partial \ln x} = \frac{\partial \ln \lambda_1}{\partial \ln x} \qquad (5.4)$$

where the second equation holds for long chains and λ_1 is the largest root of the appropriate secular equation (obtained from a matrix or from generating functions). At low density of bound protons the charged groups will on the average be far apart and titrate as if independent (like n-butylamine). Thus

(low density)
$$\xi_{1d} \to (1+x)^N,$$
$$\vartheta_{1d} \to \frac{x}{1+x},$$

(5.5)

where by low density we mean as $\vartheta \to 0$. The relations of eqn (5.5) give the limiting form of ϑ as a few protons are bound to the uncharged molecule. The other limit is where the moelcule is completely charged (all sites in the form $-NH_3^+$). Assuming that the interaction between charged groups can be described by

$$y_j = \exp(-\varphi(r_j)/RT), \qquad (5.6)$$

where $\varphi(r_j)$ is the interaction between two charged groups j units apart (since the molecule has a regular conformation r_j depends only on j and not on the position in the molecule), then for the completely charged molecule the partition function is

(saturated with H^+) $\quad \xi = (xY)^N, \qquad (5.7)$

where

$$Y = \prod_{j=1}^{\infty} y_j = \exp\left\{-\sum_{j=1}^{\infty} \varphi(r_j)/RT\right\}. \qquad (5.8)$$

If we allow the molecule to dissociate protons eqn (5.7), represented as a series in the number of protons missing, is

(high density) $\xi_{hd} \to (XY)^N (1+x^{-1}Y^{-2})^N$,

$$\vartheta_{hd} \to \frac{xY^2}{1+xY^2}. \tag{5.9}$$

The quantity ξ_{hd} in eqn (5.9) is derived as follows. The partition function for the completely charged molecule is given by eqn (5.7). Note that Y gives the interactions between a given charge and all the charges to the right (or left, it does not matter) of it; while the total energy of a given charge is the sum of interactions with charges to the left and right of it, the energy per charge in the completely charged molecule is only $\frac{1}{2}$ of this (the same $\frac{1}{2}$ that appears in eqn (2.228)). When we remove a proton we lose a factor x (hence the x^{-1} factor) and two factors Y^{-2}, since when we remove an isolated charge we lose the interactions both to the right and left of it (hence the Y^{-2} factor). The net factor $x^{-1}Y^{-2}$ gives the factors lost when a single isolated charge is removed. In the limit of few charges removed (high density of bound protons, $\vartheta \to 1$) then we can treat all missing charges as independent since on the average they will be far apart. Thus the limiting form of eqn (5.9) we have that at $\vartheta_{ld} = \frac{1}{2}$ and $\vartheta_{hd} = \frac{1}{2}$

(low density) $pH_{\frac{1}{2}} = pK$,

(5.10)

(high density) $pH_{\frac{1}{2}} = pK + 2 \ln Y$.

The titration curve for independent units (n-butylamine) is given in Fig. 5.1; also given is the titration curve for the hypothetical system described by eqn (5.9). It is straightforward, but somewhat complicated, to show that for the actual titration curve $\vartheta = \frac{1}{2}$ at a pH half way between the values for the high- and low-density titration curves given in eqn (5.10), as asymptotically approaching ϑ_{hd} and ϑ_{ld} in the limits respectively of low and high pH. Thus we have the important result

$$pK' = pK + \ln Y, \tag{5.11}$$

where pK' is the pH where $\vartheta = \frac{1}{2}$ for the actual titration curve. From Fig. 5.1 this value can be read off the graph giving $pK' = 9.1$. Thus for polylysine in 95 per cent methanol, $Y = 10^{-1}$. Since Y is the product of factors representing pair interactions (eqn (5.8)) we can now try and fit the value of Y with an appropriate pair potential.

From eqn (2.258) we have that the pair potential between two charged groups in the Debye-Hückel approximation is given by

$$\varphi(r) = \frac{1}{Dr} \exp(-\varkappa r), \qquad (5.12)$$

where K is a function of the salt concentration (eqn (2.259)). In the Debye-Hückel approximation Y is given by

$$\ln Y = -16.7 \sum_{j=1}^{\infty} \exp(-\varkappa r)/r, \qquad (5.13)$$

where the constant is $(330\,000/DRT)$ using the relative permittivity for methanol $(D = 32.6)$ and taking $T = 298K$. If it were not for the screening term $\exp(-\varkappa r)$ in eqn (5.13) the sum would diverge since

$$\sum_{j}^{\infty} 1/r \sim \int_{1}^{\infty} \frac{1}{r}\, dr = \int_{1}^{\infty} d\ln r = \infty.$$

Using eqn (2.259) for $\varkappa$ with 3.04 replaced by $3.04\,(33/80)^{\frac{1}{2}}$, where the salt concentration is $0.025\ \mathrm{mol\ l^{-1}}$, we find the predicted value of pK' given by eqn (5.11) using eqn (5.13) for Y is ~ 8.0 which is seen to be too low. If we truncate eqn (5.13) at $j = 3-4$ then we get good agreement. Or we can treat k as an empirical parameter adjusted so as to give agreement with the observed pK'; the value so obtained is $\varkappa = 1.5\ \mathrm{nm}^{-1}$, which is to be compared with the Debye–Hückel result for the case under consideration of $\varkappa = 0.8\ \mathrm{nm}^{-1}$. The study of the titration of rigid macromolecules using eqn (5.11) and eqn (5.8) gives the general relations (assuming pair interactions)

$$pK' = pK - 2.303 \sum_{j=1}^{\infty} \varphi(r_j)/RT, \qquad (5.14)$$

which promises useful information about the nature of $\varphi(r_j)$: e.g. we can study experimentally how $\varphi(r_j)$ varies with salt concentration.

While eqn (5.14) gives a connection between the pair potential $\varphi(r_j)$ and the observed midpoint of the cooperative titration of a rigid macromolecule, it does not give any further details of the curve, e.g. the slope of pH = pH'. Following the development in section 4.1, we can represent the parrition function as a matrix product where the matrix correlates a finite number of nearest-neighbour interactions (including here a finite number of repulsive interactions between neighbouring charged groups). Letting (+) and (0) represent charged and uncharged states respectively, the matrix that correlates nearest neighbour interactions is given by

$$W_2 = \begin{array}{c} + \\ 0 \end{array} \begin{bmatrix} xy_1 & x \\ 1 & 1 \end{bmatrix} \begin{array}{c} + \quad 0 \end{array} \qquad (5.15)$$

(correlation of two successive sites).

For the correlation of the ith unit with the $(i+1)$th and $(i+2)$th units we use eqn (4.36) and obtain

$$W_3 = \begin{array}{c} ++ \\ +0 \\ 0+ \\ 00 \end{array} \begin{bmatrix} xy_1y_2 & xy_1 & 0 & 0 \\ 0 & 0 & xy_2 & x \\ 1 & 1 & 0 & 0 \\ 0 & 0 & 1 & 1 \end{bmatrix} \begin{array}{cccc} +\ 0 & +\ 0 \\ +\ +\ 0\ 0 \end{array} \qquad (5.16)$$

(correlation of three successive sites).

The secular equation of the matrix given in eqn (5.15) is

$$\lambda^2 - \lambda(1+xy_1) + xy_1 - x = 0. \qquad (5.17)$$

Eqn (5.17) can be divided by λ^2 and rearranged to give

$$(1+xy_1)/\lambda + (x-xy_1)/\lambda^2 = 1. \qquad (5.18)$$

In general the secular equation of a $\rho \times \rho$ matrix can be rearranged in the form

$$G(\lambda,x,y_i) = \sum_{k=1}^{\rho} C_k(x,y_i)/\lambda^k = 1, \qquad (5.19)$$

which expresses the secular equation in the form of a generating function $G(\lambda,x,y_i)$ (see eqn (4.51)). The largest eigenvalue thus is the largest root of the equation

$$G(\lambda,x,y_i) = 1. \qquad (5.20)$$

Implicit differentiation of eqn (5.20) gives

$$\vartheta = \left(\frac{\partial \ln \lambda}{\partial \ln x}\right)_{\lambda_1} = -\left(\frac{\partial \ln G/\partial \ln x}{\partial \ln G/\partial \ln \lambda}\right)_{\lambda_1}. \qquad (5.21)$$

Letting G_m represent the generating function for the correlation of m sites then the matrices of eqns (5.15) and (5.16) yield

$$G_1 = (1+x)/\lambda,$$

$$G_2 = (1+xy_1)/\lambda + (x-xy_1)/\lambda^2,$$

$$G_3 = (1+xy_1y_2)/\lambda + (xy_2-xy_1y_2)/\lambda^2 + (x-xy_2)(1+xy_1y_2)/\lambda^3 + x^2y_1(1-y)_1/\lambda^4.$$

Computer derivation of G_m for m greater than 3 leads by induction to the leading terms in the generating function for $m = \infty$ (infinite range of correlation) [2]

$$G_\infty = (1+P_1)/\lambda + (P_2-P_1)/\lambda^2 + (P_3-P_2)(1+P_1)/\lambda^3 +$$

$$+ (P_4-P_3)(1+P_1)^2 + xP_1(1-y_2)^2/y_2/\lambda^4 + \ldots \qquad (5.22)$$

where

$$P_1 = xy,$$

$$P_k = x\prod_{j=0}^{\infty} y_{2j+k} \quad (k > 1) \qquad (5.23)$$

Using eqn (5.22) in eqn (5.21) we obtain

$$\vartheta = \frac{P_1 + (P_2-P_1)/\lambda_1 + (P_3-P_2)(1+2P_1)/\lambda_1^2 + \ldots}{(1+P_1) + 2(P_2-P_1)/\lambda_1 + 3(P_3-P_2)(1+P_1)/\lambda_1^2 + \ldots}, \quad (5.24)$$

where λ_1 is the largest root of eqn (5.22). Truncating the series in the numerator and denominator in eqn (5.24) at the leading terms yields ($P_1 = xY$)

$$\rho = \frac{xY}{1 + xY}. \quad (5.25)$$

This titration curve has the same shape as ϑ_{1d} and ϑ_{hd} (eqns (5.5) and (5.9) respectively) and lies half way between them; the titration curve given in eqn (5.25) is shown by the dot-dash curve in Fig. 5.1. As we include more terms in the series in eqn (5.24) (calculating λ_1 using the same number of terms in the series in eqn (5.22)) we obtain increasingly good estimates of the actual titration curve; we can extrapolate the behaviour of the truncated series to an infinite number of terms, the behaviour converging rapidly. One unusual property of using truncated series in eqn (5.22) and eqn (5.24) is that at all degrees of truncation we have $\vartheta = \frac{1}{2}$ at pH = pK'; this is unlike the behaviour of series expansions in the density.

While the treatment of cooperative titrations is complicated, such an analysis yields useful information about the nature of electrostatic interactions in biological macromolecules. Fortunately a great deal of information can be obtained simply from a measurement of pK' (using eqn (5.14)).

I shall conclude this section by briefly indicating how changes in the state of proton binding and changes in conformation occurring simultaneously in the same system can be treated.[3] Consider a molecule with two conformational states per unit (e.g. h and c) where each state is also capable of existing in a charged and uncharged form. Following the example of polylysine we take here the charged state to be the state with a bound proton. Then there are a total of four possible states per unit, h, h^+, c, and c^+. The matrix required to describe this system with nearest-neighbour interactions is

$$W = \begin{matrix} h \\ h^+ \\ c \\ c^+ \end{matrix} \begin{bmatrix} h & h^+ & c & c^+ \\ s & s & \sigma s & \sigma x \\ xs & xsy_{hh} & x\sigma s & x\sigma sy_{cc} \\ 1 & 1 & 1 & 1 \\ x & xy_{ch} & x & xy_{cc} \end{bmatrix} \quad (5.26)$$

where $x = [H^+]K$ and the y factors represent electrostatic repulsions between neighbouring charged groups. If the distance between the charged groups is larger for c^+c^+ than h^+h^+ then proton binding can induce a transition from helix to coil at constant temperature, the coil being favoured at high charge density since in the coil state the repulsive charges are further apart.

5.2. Binding of Mg^{2+} to DNA

The backbone of DNA contains the phosphate diester linkage

$$\text{(a)} \quad \begin{matrix} \text{OH} \\ | \\ R-O-P-O-R \\ \parallel \\ O \end{matrix} \qquad \text{(b)} \quad \begin{matrix} O^- \\ | \\ R-O-P-O-R \\ \parallel \\ O \end{matrix} \quad (5.27)$$

in each nucleotide residue. The single remaining acidic proton per phosphate group has a low pK (similar to the first pK of phosphoric acid; see eqn (1.72) and Fig. 1.8). Thus at pH ~ 7 the proton is dissociated and the form present is structure (b) in (5.27). Mg^{2+} binds strongly to phosphate groups. Designating structure (b) in (5.27) as P^-, then the binding of Mg^{2+} to isolated P^- groups is described by the reaction

$$Mg^{2+} + P^- \rightleftharpoons Mg^{2+} \cdot P^-, \quad (5.28)$$

with

$$\frac{[Mg^{2+} \cdot P^-]}{[Mg^{2+}][P^-]} = K. \quad (5.29)$$

Defining

$$[Mg^{2+}] = 10^{-pMg}, \qquad (5.30)$$

$$K = 10^{pK},$$

then the probability of the complex is given by (see eqn (1.68))

$$p(Mg^{2+} \cdot P^-) = \frac{10^{pK-pMg}}{1 + 10^{pK-pMg}} \qquad (5.31)$$

Eqn (5.31) is the exact analogue of the titration of a monoprotic acid with pMg replacing pH; thus $p(Mg^{2+} \cdot P^-) = \frac{1}{2}$ at pMg = pK, the transition taking place essentially completely over the range pMg = pK ± 1. The value of pK for the reaction given in eqn (5.28) is approximately pK = 3.78.[4]

Eqn (5.31) describes the binding of Mg^{2+} to isolated phosphate diesters. The binding of Mg^{2+} to the phosphate groups in the DNA backbone is influenced by the closeness of the binding sites. Schematically

$$\begin{array}{cccc} Mg^{2+} & Mg^{2+} & Mg^{2+} & \\ \overset{\bullet}{P^-} & \overset{\bullet}{P^-} & \rightleftharpoons \overset{\bullet}{P^-} & P^- + Mg^{2+}, \end{array} \qquad (5.32)$$

which is analogous to eqn (5.2). Thus the binding of Mg^{2+} to a site neighbouring a site with an Mg^{2+} already bound is more difficult than the case of binding to an independent site. Mathematically, the binding of Mg^{2+} to P^- in the DNA backbone is analogous to the binding of protons to polylsine discussed in the last section. We find that a very much simpler model suffices to explain the experimental binding curve.

From molecular models it appears that the existence of a hydrated Mg^{2+} ion bound to a given phosphate group sterically precludes the binding of a hydrated Mg^{2+} to nearest-neighbour phosphate groups in the polymer backbone. Letting

$$x = [Mg^{2+}]K \qquad (5.33)$$

COOPERATIVE BINDING TO BIOPOLYMERS

and using the symbols (2+) and (0) to represent phosphate groups bound and unbound with Mg^{2+} then the matrix that excludes nearest-neighbour occupancy is

$$W = \begin{array}{c} \\ 2+ \\ 0 \end{array} \begin{array}{cc} 2+ & 0 \\ \left[\begin{array}{cc} 0 & x \\ 1 & 1 \end{array} \right] \end{array} \qquad (5.34)$$

The largest eigenvalue of the matrix in eqn (5.34) is

$$\lambda_1 = \tfrac{1}{2}\{1 + (1+4x)^{\tfrac{1}{2}}\}. \qquad (5.35)$$

As usual the fraction, ϑ, of bound sites is given by (for long chains)

$$\vartheta = \frac{\partial \ln \lambda_1}{\partial \ln x}. \qquad (5.36)$$

Using eqn (5.35) we have

(independent sites) $\qquad \vartheta = \dfrac{x}{1 + x} \qquad (5.37a)$

$\begin{pmatrix}\text{nearest-neighbour occupancy}\\ \text{of forbidden sites}\end{pmatrix} \quad \vartheta = \dfrac{2x}{1 + 4x + \sqrt{(1+4x)}}. \,(5.37b)$

Note that eqn (5.37b) contains no unknown parameters. Note further that, for large x (large $[Mg^{2+}]$),

(independent sites) $\qquad\qquad \vartheta \to 1,$

$\begin{pmatrix}\text{nearest-neighbour occupancy}\\ \text{of forbidden sites}\end{pmatrix} \quad \vartheta \to \tfrac{1}{2}$
$\qquad (5.38)$

Fig. 5.2 illustrates ϑ as calculated from eqns (5.37a) and (5.37b); also shown is binding data for DNA double helix.[4] We see that eqn (5.37b), where nearest-neighbour occupancy of sites is forbidden (leading in the limit of $\vartheta \to \tfrac{1}{2}$ to the state of binding where every other site is occupied), adequately fits the experimental binding data for DNA. Similar data[4] for single strands, poly(A—T) double helix, and poly(A) double helix (which is not the Watson—Crick helix) is almost superimposable

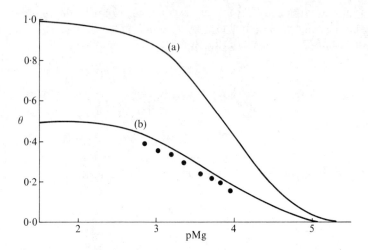

FIG. 5.2. Mg^{2+} binding to DNA. Curve (a) gives the titration curve for independent phosphate groups; curve (b) gives the calculated curve for the model described in the text; the solid points are experimental data for native double-strand DNA.

on that shown for double-strand DNA.

5.3. Cooperative binding to surfaces

A great deal is known about the statistical mechanics of particles on regular lattices. The model is a regular two-dimensional lattice of sites, where a site can be occupied or unoccupied with interactions between nearest-neighbour occupied sites. Two lattices, the simple square and honeycomb lattices, are illustrated in Fig. 5.3, where the solid dots represent occupied sites, the dotted lines indicating interactions between nearest-neighbour occupied sites. If the binding constant for the reaction

$$B + \text{sites} \rightleftharpoons B\cdot\text{site} \tag{5.39}$$

is K (B a general species that binds to the surface) then

$$x = [B]K \tag{5.40}$$

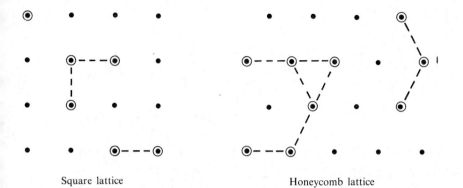

FIG. 5.3. Illustration of two two-dimensional lattices. The dotted lines indicate interactions between sites occupied by particles.

is the statistical weight for binding to an isolated site, where [B] is the concentration of B free in solution in equilibrium with the surface. The quantity

$$y = \exp(-\varepsilon/RT) \qquad (5.41)$$

is the statistical weight representing the interaction between neighbouring occupied sites (ε the energy or free energy of interaction). The average partition function per site λ_1 is given as a series in x:

$$\ln \lambda_1 = \sum_{n=1}^{\infty} b_n(y) \, x^n, \qquad (5.42)$$

where the fraction of occupied sites is given in the usual fashion

$$\vartheta = \frac{\partial \ln \lambda_1}{\partial \ln x} = \sum_{n=1}^{\infty} n b_n(y) \, x^n, \qquad (5.43)$$

where

$$\begin{pmatrix}\text{square}\\\text{lattice}\end{pmatrix} \quad \begin{aligned} b_1 &= 1, \\ b_2 &= -2(\tfrac{1}{2}) + 2y, \\ b_3 &= 10(\tfrac{1}{3}) - 16y + 6y^2, \\ b_4 &= -52(\tfrac{1}{4}) + 118y - 85y^2 + 18y^3 + y^4, \end{aligned} \qquad (5.44)$$

and

$$\begin{pmatrix}\text{honeycomb}\\\text{lattice}\end{pmatrix}\quad\begin{array}{l}b_1 = 1\\ b_2 = -3\tfrac{1}{2} + 3y,\\ b_3 = 19\tfrac{1}{3} - 30y + 9y^2 + 2y^3,\\ b_4 = 129\tfrac{3}{4} + 288y - 178\tfrac{1}{2}\,y^2 + 5y^3 + 12y^4 + 3y^5.\end{array}\quad(5.45)$$

Eqns (5.44) and (5.45) give the first few terms in eqn (5.42); the terms up to b_{13} are known for many lattices.[5] Since eqn (5.42) is a series in x, $x = [B]k$, the truncated series gives ϑ accurately at low $[B]$. For the model of Fig. 5.3, the high-density series is given in terms of the same b_n as given in eqns (5.44) and (5.45),

$$\ln \lambda_1 = \ln xy^2 + \sum_{n=1}^{\infty} b_n(y)\, z^n \quad (z = x^{-1}y^{-4}) \quad (5.46)$$

(square lattice)

and

$$\ln \lambda_1 = \ln xy^3 + \sum_{n=1}^{\infty} b_n(y)\, z^n \quad (z = x^{-1}y^{-6}) \quad (5.47)$$

(honeycomb lattice).

Truncated series using the b_n in eqns (5.44) and (5.45) for the quantities $\ln \lambda_1$ given in eqns (5.46) and (5.47) give ϑ as $\vartheta \to 1$ via the first equation in (5.43).

The models of Fig. 5.3 show a first-order phase transition and a critical point. By studying binding to surfaces in the limits of low and high degrees of occupancy, we can determine y and K by matching theory with experiment.

For independent sites ($y = 1$)

$$\lambda_1 = 1 + x, \quad (5.48a)$$

$$\vartheta = \frac{x}{1+x} = \frac{[B]K}{1+[B]K} \quad (5.48b)$$

Eqn (5.48b) is known as the Langmuir adsorption isotherm

although it is seen to be identical with the equation for the titration of a monoprotic acid (eqn (1.50)). Fig. 5.4 shows binding to the square lattice as a function of pB = $-\ln[B]$ comparing eqn (5.48b) ($y = 1$) with the series expansions. The portions of the curve labelled 'high' and 'low' respectively are given by eqn (5.46) and eqn (5.42) respectively. The case illustrates is for ε negative (attractive interaction). For the value of y illustrated there is a discontinuity (first-order phase transition) in the binding isotherm; as the temperature is increased (y thus decreasing toward unity) the size of the discontinuity decreases until at y_c, the critical temperature, the discontinuity disappears and the binding isotherm is a continuous sigmoidal curve.

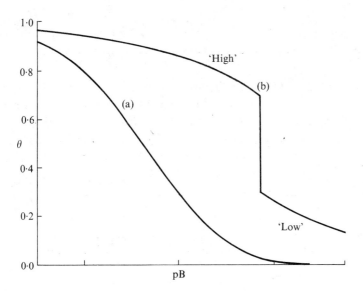

FIG. 5.4. Illustration of binding of particles to a two-dimensional lattice. Curve (a) is for independent sites (Langmuir adsorption isotherm) while curve (b) is for the cooperative model with interactions between neighbouring occupied sites illustrating a phase transition.

5.4. Cooperative binding of oligomers to polymers

The synthetic polynucleotides polyinosinate (poly(I)) and polycytidylate (poly(C)) associate to give a double-strand complex poly(I)·poly(C) forming double helix.[6] In like manner oligomers of inosine I_n also bind to poly(C), forming double helix (n the number of bases in the oligomer). The net reaction b can be schematically described as

$$C_N + (N/n)I_n \rightleftharpoons C_N \cdot (I_n)_{N/n}, \tag{5.49}$$

with all intermediate states of binding possible:

$$C_N + I_n \rightleftharpoons C_N \cdot (I_n),$$
$$C_N + 2\,I_n \rightleftharpoons C_N \cdot (I_n)_2. \tag{5.50}$$

In this section we shall use the poly(C)·I_n association as an example of the binding of intermediate-size molecules to large molecules.

Fig. 5.5 illustrates schematically the two parameters required to treat the binding. In reaction (I) an oligomer binds to the polymer at a given location on the polymer free of bound oligomer with equilibrium constant K_n, the constant being a

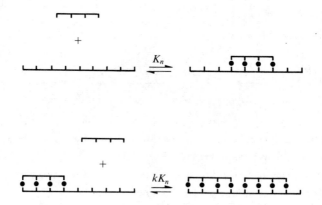

FIG. 5.5. Illustration of the parameters required to describe the all-or-none binding of oligomers to polymers.

function of oligomer size n. In reaction (II) an oligomer binds next to an oligomer that is already bound; the intrinsic binding constant K_n is modified with an interaction parameter k that is independent of oligomer size. As shown in Fig. 5.5 we assume that the only state of binding of the oligomer is when the maximum amount possible of double helix between oligomer and polymer is formed (all-or-none binding). We have already seen in section 3.1 that short oligomers associate mainly in the all-or-none fashion.

The starting pint and one of the most important parts of any association equilibrium are the conservation relations. We define

N = number of bases per chain in the poly(C),

n = number of bases per chain in I_n,

$[C_N]_0$ = initial concentration of poly(C), (5.51)

c = $[I_n]_0$ = initial concentration of I_n,

a = $[I_n]$ = concentration of free I_n at equilibrium.

We treat the case where poly(C) and I_n are mixed with a one-to-one ratio between the total number of bases from each species in the total system

$$c = [I_n]_0 = (N/n)[C_N]_0. \quad (5.52)$$

Defining the fractionof helix (taken here as the fraction of bases paired) as

$$\vartheta = \text{fraction of helix} = \text{fraction of oligomers bound}, \quad (5.53)$$

where the second equality follows only for all-or-none binding. Using eqns (5.51)–(5.53) the conservation of oligomers is given by

$$a + c\vartheta = c \quad \text{(conservation of oligomers)}. \quad (5.54)$$

We showed in section 3.1 that the study of the variation of

ϑ with concentration was the simplest approach to association equilibrium. We use that approach here and define

$$\rho = \left(\frac{\partial \ln \vartheta}{\partial \ln c}\right)_T. \qquad (5.55)$$

We then have the following general relations

$$\vartheta = \vartheta(a, k, K_n),$$
$$p = \rho(a, k, K_n), \qquad (5.56)$$
$$a + c\vartheta(a, k, K_n) = c$$

If ϑ and ρ are determined experimentally, then eqn (5.56) gives three equations in the three unknowns a, k, and K_n. I shall now give specific relations for the general equations given in eqn (5.56).[7]

If ξ is the molecule grand partition function (sum over all state of binding of I_n to poly(C)) then the average number of oligomers bound to a polymer is given by

$$\langle M \rangle = \frac{\partial \ln \xi}{\partial \ln a} = \frac{\partial \ln \xi}{\partial \ln K_n} \qquad (5.57)$$

where the first equality holds in general and the second only for all-or-none binding. The fraction of helix (eqn (5.53)) is given using eqn (5.57) as

$$\vartheta = \frac{n\langle M \rangle}{N} = \frac{n}{N} \frac{\partial \ln \xi}{\partial \ln K_n}. \qquad (5.58)$$

To formulate a relation for ρ (eqn (5.55)), we note that

$$\frac{\partial \ln \vartheta}{\partial \ln c} = \frac{\partial \ln \vartheta}{\partial \ln a} \frac{\partial \ln a}{\partial \ln c} = \frac{\partial \ln \vartheta}{\partial \ln K_n} \frac{\partial \ln a}{\partial \ln c}, \qquad (5.59)$$

where the second equality holds again only for all-or-none binding. Differentiation of eqns (5.58) and (5.54) yields

$$\frac{\partial \ln \vartheta}{\partial \ln K_n} = \frac{n}{N\vartheta} \frac{\partial^2 \ln \xi}{\partial (\ln K_n)^2} \qquad (5.60)$$

and

$$\frac{\partial \ln a}{\partial \ln c} = 1 - \left(\frac{\vartheta}{1-\vartheta}\right)\frac{\partial \ln \vartheta}{\partial \ln c}. \tag{5.61}$$

Using eqns (5.60) and (5.61) in eqn (5.59) gives a relation for ρ:

$$\rho = 1 \Big/ \left\{ \frac{\vartheta}{1-\vartheta} + \left(\frac{n}{N\vartheta}\,\frac{\partial \ln \xi}{\partial \ln K_n^2}\right)^{-1} \right\}$$

$$= 1 \Big/ \left\{ \frac{\vartheta}{1-\vartheta} + \left(\frac{\partial \ln \vartheta}{\partial \ln K_n}\right)\right\}^{-1}. \tag{5.62}$$

For the all-or-none binding of oligomers to polymers eqns (5.58) and (5.62) are general relations. I shall now give an explicit formulation of ξ.

We first treat the limiting case of infinite poly(C) (I_n of course being of finite length). The polymer can be viewed as having alternating runs of coil (no oligomer bound) and runs of oligomers in consecutive sequence (isolated oligomer are sequences containing only one unit). The statistical weights of these sequences are

$$u(h_j) = (K_n k)^j k^{-1}, \tag{5.63a}$$

$$u(c_j) = 1. \tag{5.63b}$$

The coil statistical weight is unity (eqn (5.63b)) since as usual we take the coil state as the reference state. The statistical weight for helical sequences given in eqn (5.63a) arises from the fact that every oligomer bound has a binding constant K_n; for j oligomers in a consecutive sequence there are $(j-1)$ interaction parameters k corresponding to the $(j-1)$ interfaces between oligomers. Using eqn (5.63) in eqn (4.50) we have

$$U_c(\lambda) = \sum_{j=1}^{\infty} (1/\lambda)^j = \frac{1}{\lambda - 1}, \tag{5.64a}$$

$$U_h(\lambda) = \sigma \sum_{j=1}^{\infty} \left(\frac{r}{\lambda^n}\right)^j = \frac{\sigma r}{\lambda^n - r}, \tag{5.64b}$$

$$U_c(\lambda)\, U_h(\lambda) = 1, \tag{5.64c}$$

where for simplification we define

$$r = akK_n. \tag{5.65}$$

Note that λ_1, the largest root of eqn (5.64c), is the partition function per unit; hence in eqn (5.64b) a run of j oligomers in sequence spans jn nucleotide residues in poly(C) (thus the factor λ^{-n} for each oligomer). Explicitly eqn (5.64c) is

$$\lambda^{n+1} - \lambda^n - \lambda r + (1-\sigma)r = 0. \tag{5.66}$$

For long chains eqn (5.58) is given by

$$\vartheta = n \frac{\partial \ln \lambda_1}{\partial \ln r}. \tag{5.67}$$

Implicit differentiation of eqn (5.66) yields the derivative

$$\frac{\partial \ln \lambda_1}{\partial \ln r} = \frac{r(\lambda_1 - 1 + \sigma)}{(n+1)\lambda_1^{n+1} - n\lambda_1^n - \lambda_1 r}. \tag{5.68}$$

Solving eqn (5.66) for λ_1^n/r we have

$$\lambda_1^n/r = (\lambda_1 - 1 + \sigma)/(\lambda_1 - 1),$$

and subsitution of the above result into eqn (5.68) yields

$$\vartheta = 1/1 + \frac{\sigma \lambda_1}{n(\lambda_1 - 1)(\lambda_1 - 1 + \sigma)}. \tag{5.69}$$

Using eqn (5.69) the general relation for (eqn (5.62)) can be calculated using the relation

$$\frac{\partial \ln \vartheta}{\partial \ln r} = \frac{\partial \ln \vartheta}{\partial \ln \lambda_1} \frac{\partial \ln \lambda_1}{\partial \ln r},$$

giving

$$\rho = \frac{1}{\vartheta/(1-\vartheta) + (1-\vartheta)^2 \{2(\lambda_1 - 1)/\sigma + 1\} - \vartheta(1-\vartheta)/n}. \tag{5.70}$$

From an experimental measurement of ρ at a given value of ϑ eqns (5.69) and (5.70) represent two equations in two unknowns, σ and λ_1. The equations are seen to be quadratic in these variables and hence we can obtain explicit solutions. Knowing σ and λ_1, r can be calculated using eqn (5.66). Notice that although eqn (5.66) is an $(n+1)$th-order polynomial in λ, by the above procedure we can determine the parameters r and σ from a single measurement of ρ exactly without having to obtain the roots of eqn (5.66). The reader might want to review the derivation and note just how solving the secular equation (eqn (5.66)) was avoided (implicit differentiation of eqn (5.66) and then use of eqn (5.66) to eliminate high powers of λ_1).

We can produce even simpler results by introducing the average length (in terms of the number of base pairs) of a helical sequence (see eqn (4.103))

$$\langle L \rangle = n \frac{\partial \ln \xi}{\partial \ln r} / \frac{\partial \ln \xi}{\partial \ln \sigma}$$

$$= n \frac{\partial \ln \lambda_1}{\partial \ln r} / \frac{\partial \ln \lambda_1}{\partial \ln \sigma}. \qquad (5.71)$$

Using the same procedure used to obtain eqn (5.68) we have

$$\langle L \rangle = \left(\frac{\vartheta}{1-\vartheta}\right)\left(\frac{\lambda_1}{\lambda_1 - 1}\right). \qquad (5.72)$$

Eqns (5.70)–(5.72) give

$$\langle L \rangle = \tfrac{1}{2}\left(\frac{n\rho}{(1-\vartheta)^2 - \rho\vartheta(1-\vartheta)} + n + \frac{\vartheta}{1-\vartheta}\right),$$

$$k = [\{(1-\vartheta)/\vartheta\}\langle L \rangle^2 - \langle L \rangle\{n(1-\vartheta)/\vartheta + 1\} + n]/n \qquad (5.73)$$

$$K_n = \frac{n\vartheta}{c(1-\vartheta)^2 \langle L \rangle^2}\left\{\frac{\langle L \rangle}{\langle L \rangle - \vartheta/(1-\vartheta)}\right\}^{n+1}.$$

The most useful value of ρ is at $\vartheta = \tfrac{1}{2}$. Designating quantities referring to $\vartheta = \tfrac{1}{2}$ by the subscript m (ρ_m, $\langle L \rangle_m$, T_m), eqn (5.73) becomes

$$\langle L \rangle_m = \frac{\rho_m(3n-1) + n + 1}{2(1-\rho_m)},$$

$$k = \{\langle L \rangle_m^2 - (n+1)\langle L \rangle_m + n\}/n, \tag{5.74}$$

$$K_n = \frac{2n}{c\langle L \rangle_m^2} \left[\frac{\langle L \rangle_m}{\langle L \rangle_m - 1}\right]^{n+1}.$$

In eqn (5.56) I indicated that a measurement of ρ at a given value of ϑ gave three equations in three unknowns; eqn (5.74) gives the constants k and K_n explicitly in terms of the experimental qualities ρ and ϑ with the intermediate parameter $\langle L \rangle$ (of physical interest in itself).

The case of $n = 1$ and $k = 1$ is a useful reference case. These conditions correspond to binding to independent sites (Langmuir adsorption isotherm — eqn (5.48b)); we have at $\vartheta = \frac{1}{2}$: $\rho_m = \frac{1}{3}$, $\langle L \rangle_m = 2$, $k = 2/c$.

We can use eqn (5.73) to give ρ_m as a function of the interaction parameter k only. In this manner it is easy to show that $\rho_m(n,k)$ has the following behaviour,

$$\ln k \to \infty, \quad \rho_m \to 1, \quad \langle L \rangle_m \to \infty \quad \text{(all } n\text{)},$$
$$\ln k = 0, \quad \rho_m = \left(\frac{n+1}{5n+1}\right), \quad \langle L \rangle_m = n+1, \tag{5.75}$$
$$\ln k \to -\infty, \quad \rho_m \to \left(\frac{n-1}{5n-1}\right), \quad \langle L \rangle_m \to n.$$

Thus we have

$$1 > \rho_m > \left(\frac{n-1}{5n-1}\right). \tag{5.76}$$

The behaviour of ρ_m as a function of n and $\ln k$ is shown in Fig. 5.6. We see from Fig. 5.6 that only for a limited range of $\ln k$ is $\rho_m(n,k)$ sensitive to this quantity, ρ_m approaching the asymptotic limits given in eqn (5.76) for $\ln k \gg 0$, of $\ln k \ll 0$. The parameter k plays the role of a cooperativity parameter, the parameter σ in this model being $\sigma = k^{-1}$. The relation of k to the cooperativity and the behaviour of ρ_m is

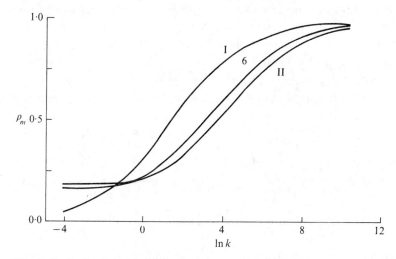

FIG. 5.6. Variation of the quantity ρ_m with $\ln k$.

given below

k	oligomer–oligomer interaction	cooperativity
$k > 1$ ($\ln k > 0$)	attractive (oligomers tend to cluster)	positive
$k = 1$ ($\ln k = 0$)	neutral	independent
$k < 1$ ($\ln k < 0$)	repulsive (oligomers tend to be isolated)	negative.

From Fig. 5.6, we see that for very positive or very negative cooperativity, ρ_m is independent of the precise value of k.

The curves shown in Fig. 5.6 are calculated most conveniently from the relations (derived from eqn (5.74))

$$\rho_m = \tan h\{\tfrac{1}{2}\ln[\langle L\rangle_m/n + (n-1)/2n]\},$$

(5.77)

$$\langle L\rangle_m = \tfrac{1}{2}[n + 1 + \{(n-1)^2 + 4nk\}^{\tfrac{1}{2}}].$$

Details of the application of the above treatment (including the influence of finiteness of the poly(C) to experimental data for the poly(C)–I_n systems are given in the literature.[7] I shall quote here results that illustrate the nature of the cooperative equilibrium.

The procedure in outline is as follows. Using the experimental value of ρ_m, we pick off the value of $\ln k$ from Fig. 5.6. Fortunately the experimental values of ρ_m for the poly(C)–I_n system are in the neighbourhood of $\rho_m \sim \frac{1}{2}$, the most sensitive part of the graph. Using values of ρ_m measured at different temperatures gives $\ln k$ as a function of temperature. Then the thermodynamic parameters for k are given as

$$\frac{\partial \ln k}{\partial (1/T)} = -\frac{\Delta h_k}{R},$$

$$\Delta g_k = -RT \ln k, \qquad (5.78)$$

$$\Delta s_k = \frac{\Delta h_k - \Delta g_k}{T}.$$

The values so obtained are

$$\Delta h_k = -43.5 \pm 8.8 \text{ kJ mol}^{-1}$$

$$\Delta s_k = -110 \pm 29 \text{ J mol}^{-1} \text{ k}^{-1} \qquad (5.79)$$

Table 5.1 shows how k varies as a function of temperature. We see that since $k > 1$ for all temperatures in the range 273K to 373K, the binding of I_n is cooperative, the oligomers tending to cluster. A measure of this clustering tendency is $\langle L \rangle_m$.

TABLE 5.1

Variation with temperature of the oligomer-oligomer interaction constant k and the average cluster size m, at $\vartheta = \frac{1}{2}$ for n = 6 and n = 10 for for infinite polymers

T(K)	k	$\langle m \rangle$ n = 6	n = 10
283	201	6.4	5.1
293	108	4.8	3.9
313	34.6	3.0	2.5
333	12.8	2.1	1.8
353	5.26	1.6	1.4
373	2.39	1.3	1.2

The quantity

$$\langle m \rangle = \langle L \rangle_m / n, \qquad (5.80)$$

which is the average number of oligomers in a sequence, is also shown in Table 5.1. Note that all the $\langle m \rangle$ in Table 5.1 are for $\vartheta = \frac{1}{2}$, the temperature where $\vartheta = \frac{1}{2}$ is variable by changing the concentration. We note that the clustering tendency drops markedly as the temperature increases.

Since $\sigma = k^{-1}$, knowing k we can calculate λ_1 from eqn (5.69). Knowing λ_1 and σ, the quantity r can be calculated from eqn (5.66). Since eqn (5.54) gives at $\vartheta = \frac{1}{2}$ ($a = \frac{1}{2}$), eqn (5.65) then gives K_n. We can determine K_n at various temperatures, thus obtaining the thermodynamic parameters (as in eqn (5.78)) as a function of n. We find that

$$K_n = \exp(-\Delta h_n / RT + \Delta s_n / R),$$

$$\Delta h_n = \Delta h_0 + n \Delta h_u, \qquad (5.81)$$

$$\Delta s_n = \Delta s_0 + n \Delta s_u.$$

Eqn (5.85) gives the result that Δg_n is linear in n with a correction for the ends. This is the same form as we assumed in eqn (4.95). Note that here we have determined Δh_n and Δs_n independently for each n, and that eqn (5.81) is an experimental result. The parameters in eqn (5.81) are[7]

$$\Delta h_0 = +57.3 \pm 18.4, \qquad \Delta h_u = -27.8 \pm 2.1 \text{ kJ mol}^{-1}$$

$$\Delta s_0 = 136 \pm 29, \qquad \Delta s_u = -79 \pm 6 \text{ J mol}^{-1} \text{ K}^{-1}. \qquad (5.82)$$

Note that we need not assume that the enthalpies and entropies are independent of temperature (eqn (5.78)), although in the present case they are constant over the range of temperature studied.

Using the experimentally determined form of eqn (5.81) we can write

where

$$\beta = \exp(-\Delta h_0/RT + \Delta s_0/r),$$

$$s = \exp(-\Delta h_u/RT + \Delta s_u/R) = \exp(\Delta h_u(T-T_0)/RT\,T_0), \quad (5.83)$$

$$T_0 = \frac{\Delta h_u}{\Delta s_u} = 354 \text{ K}.$$

$$K_n = \beta s^n,$$

Table 5.2 shows β, s, and K_6 and K_{10} as a function of temperature. We see that these factors vary dramatically with temperature.

TABLE 5.2

Variation with temperature of the oligomer-polymer binding constants K_n for n = 6 and n = 10 and the parameters β and s

(K)	β	s	K_6	K_{10}
273	1.48×10^{-4}	16.22	2 690	1.86×10^8
293	8.13×10^{-4}	6.92	89.2	2.04×10^5
313	3.63×10^{-3}	3.39	5.50	724
338	1.38×10^{-2}	1.78	0.23	4.36
353	4.37×10^{-2}	1.00	0.044	0.044
373	1.23×10^{-1}	0.60	0.006	0.0008

A specific interpretation of the enthalpy Δh_n in eqn (5.81) is in terms of stacking interactions and base pair hydrogen bonding (see section 3.2.3). Since Δh_n is the enthalpy for binding an isolated oligomer there will be n base-pair hydrogen-bonding interactions and $(n-1)$ stacing interactions. Assuming that these are the only contributions to Δh_n we can write

$$\Delta h_n = n\Delta h_{HB} + (N-1)\Delta h_{ST} = -\Delta h_{ST} + n(\Delta h_{HB} + \Delta h_{ST}), \quad (5.84)$$

where Δh_{HB} and Δh_{ST} are the contributions due to hydrogen bonding between base pairs and stacking respectively. Comparing eqn (5.81) and eqn (5.82) with eqn (5.84) we find

$$\Delta h_{ST} = -57.3 \pm 18.4 \text{ kJ mol}^{-1},$$
$$\Delta h_{HB} = +29.3 \pm 16.7 \text{ kJ mol}^{-1}.$$
(5.85)

While the error is large, Δh_{HB} is definitely positive in this interpretation of Δh_n. Cytosine has three sites for hydrogen bonding; only two are used in the C—I interaction. Thus it is reasonble that Δh_{HB} is positive since one site on cytosine, when it binds to inosine, is not permitted to hydrogen bond to anything (it can hydrogen-bond to water in the coil state).

The probability distributions of helical sequence lengths given in eqn (4.112) for polypeptides can be put in a useful form using eqn (5.72) and eqn (5.80), giving

(Probability a helical sequence contains m oligomers)

$$p_m = (1 - 1/\langle m \rangle)^m (\langle m \rangle - 1)$$
(5.86)

$\begin{pmatrix}\text{Probability an oligomer is in a helical sequence containing } m \\ \text{oligomers}\end{pmatrix}$

$$P_m = m(1 - 1/\langle m \rangle)^{m+1} (\langle m \rangle - 1)^2.$$
(5.87)

The distributions p_m and P_m are shown in Figs. 5.7 and 5.8 respectively for $n = 10$. Once again, each point on the curve is for a concentration such that $\vartheta = \frac{1}{2}$. Thus the nature of the clustering of oligomers at $\vartheta = \frac{1}{2}$ varies markedly with temperature.

5.5. Summary

In this chapter I have given several examples of cooperative binding to macromolecules in detail. In general the analysis of cooperative binding is more complex than cooperative conformational transitions not involving binding, e.g. the helix—coil transition in polypeptides. While the analysis is more complicated, we have an experimental variable — concentration — that

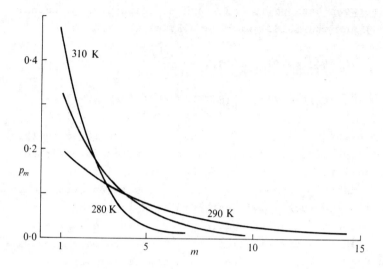

FIG. 5.7. Probability that a helical sequence contains m oligomers as a function of temperature (all for $\vartheta = \frac{1}{2}$).

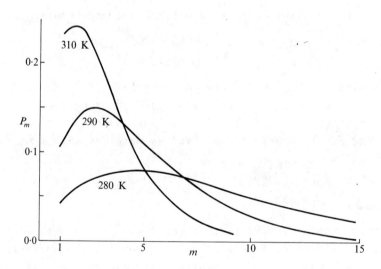

FIG. 5.8. Probability that a bound oligomer is in a sequence of m oligomers as a function of temperature (all for $\vartheta = \frac{1}{2}$).

is not available in conformational transitions alone. The use of concentration as an experimental variable, as we have seen in the examples just given, greatly reduces the number of assumptions we must make in formulating the model, and ultimately, although the derivations are more complex, leads to very simple relations (e.g. eqn (5.74)) for the parameters of the model in terms of measurable numbers.

Notes and References

1. Liem, R.K.H., Poland, D., and Scheraga, H.A., *J. Am. chem. Soc.* **92**, 5717 (1970).
2. Poland, D., *J. chem. Phys.* **58**, 3574 (1973).
3. See P&S, Chapter 6. Also: Zimm, B.H. and Rice, S.A., *Molec. Phys.* **3**, 391 (1960).
4. Sander, C. and Ts'o, P.O.P., *J. molec. Biol.* **55**, 1 (1970).
5. Sykes, M.E., Essam, J.W., and Gaunt, D.S., *J. math. Phys.* **6**, 283 (1965).
6. Tazawa, I., Tazawa, S., and Ts'o, P.O.P., *J. molec. Biol.* **66**, 115 (1972).
7. Springgate, M. and Poland D., *Biopolymers* **12**, 2241 (1973).

6
COOPERATIVE EQUILIBRIA IN SPECIFIC SEQUENCE MACROMOLECULES

Proteins and nucleic acids are sepecific-sequence molecules, the possible units in a specific sequence being respectively the approximately 20 natural amino acids and the four natural bases. Since the different units that make up a specific sequence will in general have different conformational propensities, we must take into account the nature of a unit at a given position in the molecule in formulating the partition function. For example, in Chapter 4 we showed that for homopolyamino acids with a chain length $N < 2\sigma^{\frac{1}{2}}$ the partition function can be approximated by considering only a single helical sequence of any length on the molecule

$$q \simeq 1 + \sigma \sum_{n=1}^{N} (N-n+1)s^n. \tag{6.1}$$

Using the same approximation (single helical sequence) for a specific-sequence polyamino acid we must take into account the fact that given a helical sequence of n units, we can no longer assign a single factor s^n to all such sequences in the molecule but that the statistical weight of a sequence of n units depends on its position in the chain. The partition function of eqn (6.1) for a specific sequence molecule is given by

$$q = 1 + \sum_{i=1}^{N} \sum_{j=i}^{N} \sigma(i,j) \prod_{k=i}^{j} s_k, \tag{6.2}$$

where $\sigma(i,j)$ and s_k are functions of the position in the sequence. For an arbitrary sequence it is clear that it is very difficult to deal with partition functions like eqn (6.1), the only recourse being numerical analysis.

In this chapter we will consider three questions. First I will outline a general technique for treating specific-sequence molecules. Since, of necessity, the treatment of specific-sequence molecules requires numerical analysis, the technique we will discuss is designed from the start to utilize a minimum of numerical analysis. This technique avoids cumbersome expressions like eqn (6.2), yet is very general. In fact we will show that using this approach we can treat systems having heterogeneous helix propensity (sequence-dependent ss and σs), intermediate-range specific-sequence electrostatic effects, long-range entropy effects (loops), a difference between unwinding from the ends and the interior, and chain dissociation exactly (no approximations to the model assumed). And the technique does all of the above with an absolute minimum of computer programming and time. Second, we will discuss some general features of how the presence of a specific sequence alters the nature of conformational equilibria. In particular we will examine the tendency for helical sequences to be localized in the sequence and what effect specific sequence has on the distribution of helix sequence lengths. Finally we will examine examples of specific-sequence molecules including globular proteins, collagen, and DNA.

6.1. Probability profiles; nearest-neighbour interactions

Since in specific-sequence molecules there will be regions in the sequence that more strongly favour helix than others, it is of interest to examine the probability that a particular unit in the sequence is in the helical state. The probability of being in the helix state as a function of position in the sequence is known as the probability profile. The probability profile is given in principle as follows.[2] Consider the two sequences

$$(\text{I}) \quad \underset{\text{x x x x x h x x x} \ldots \text{x,}}{1 \qquad\qquad m \qquad\qquad N}$$

$$(\text{II}) \quad \underset{\text{x x x x x x x x x} \ldots \text{x,}}{1 \qquad\qquad m \qquad\qquad N} \tag{6.3}$$

where the xs mean that the corresponding units can be in either the helix or coil states; in sequence (I) the mth unit can only be in the helix state. Then the probability that the mth unit indeed is in the helix state is given by

$$p(h_m) = q_I/q_{II}, \tag{6.4}$$

where the qs are the partition functions for the respective sequences in (6.3). For a model where the partition function can be expressed as a matrix product, we have

$$q_{II} = \mathbf{e}\left(\prod_{n=1}^{N} \mathbf{W}_{k(n)}\right)\mathbf{e}^+,$$

$$q_I = \mathbf{e}\left(\prod_{n=1}^{m-1} \mathbf{W}_{k(n)}\right)\mathbf{W}'_{k(m)}\left(\prod_{n=m+1}^{N} \mathbf{W}_{k(n)}\right)\mathbf{e}^+,$$

(6.5)

where the elements of $\mathbf{W}_{k(n)}$ depend on the nature of the unit at site n in the sequence. $\mathbf{W}'_{k\ m}$ is a matrix that allows the mth unit to be in the helix state only. For example, for the nearest-neighbour polypeptide model (s and σ being functions of position in the sequence),

$$\mathbf{W} = \begin{bmatrix} s & \sigma s \\ 1 & 1 \end{bmatrix}$$

$$\mathbf{W}' = \partial \mathbf{W}/\partial \ln s = \begin{bmatrix} s & \sigma s \\ 0 & 0 \end{bmatrix}$$

(6.6)

We see in eqn (6.6) that the matrix elements in $\mathbf{W}'$ corresponding to the coil state are set equal to zero. The evaluation of $p(h_m)$ by computer using eqns (6.4) and (6.5) is relatively simple. Shortly, we will apply eqn (6.5) to treat specific-sequence electrostatic effects in polypeptides.

For the model of eqn (6.6) with the specific sequence determining the values of σ and s, Lacombe and Simha[2] have given a simpler method of calculating $p(h_m)$ then eqn (6.5). I present here a simplified version of the approach of Lacombe and Simha.[3] Consider the two sequences with the states of the units shown fixed:

$$(I) \quad \ldots \quad h \overset{N}{h},$$
$$(II) \quad \ldots \quad h \overset{N}{c}.$$

(6.7)

The conditional probability $P(h_{N-1}|h)$ that given that the $(N-1)$th unit is h and the Nth unit is h is given by

$$P(h_{N-1}|h) = \frac{q_I}{q_I+q_{II}}, \qquad (6.8)$$

where the qs are the partition functions for the respective sequences shown in (6.7). The usefulness of eqn (6.8) lies in the fact that the qs in eqn (6.8) can be written as (as in the matrix of eqn (6.6) we arbitrarily assign the factor σ to an hc or h-end boundary)

$$q_I = qs_{N-1}s_N\sigma,$$
$$q_{II} = qs_{N-1}\sigma, \qquad (6.9)$$

where q is the partition function for the rest of the molecule (not shown as fixed in (6.7)). To simplify the presentation of this approach we take σ as independent of sequence (it is not difficult to include sequence-dependent σs). Using eqn (6.9) in eqn (6.8) we have

$$P(h_{N-1}|h) = \frac{s_N}{1+s_N}. \qquad (6.10)$$

By consideration of the sequences

$$\begin{array}{ll}(I) & \ldots c \overset{N}{c}, \\ (II) & \ldots c \overset{N}{h}, \end{array} \qquad (6.11)$$

we have

$$P(c_{N-1}|c) = \frac{q_I}{q_I+q_{II}}, \qquad (6.12)$$

with

$$q_I = qs_N\sigma,$$
$$q_{II} = q, \qquad (6.13)$$

giving

$$P(c_{N-1}|c) = \frac{1}{1+\sigma s_N}. \tag{6.14}$$

Using the general conservation relations for the conditional probabilities

$$P(h_m|h) + P(h_m|c) = 1,$$
$$P(c_m|c) + P(c_m|h) = 1, \tag{6.15}$$

the other two conditional probabilities can be calculated. For example,

$$P(c_{N-1}|c) = 1 - P(c_{N-1}|h). \tag{6.16}$$

Now consider the sequences

$$\begin{array}{l} \overset{m}{} \overset{N}{} \\ (I) \quad \ldots \text{h c h h h h h}, \\ \overset{m}{} \overset{N}{} \\ (II) \quad \ldots \text{h h h h h h h}. \end{array} \tag{6.17}$$

The ratio of the *a priori* probabilities of the two sequences shown is

$$\frac{p(I)}{p(II)} = \frac{q_I}{q_{II}}, \tag{6.18}$$

where

$$\begin{aligned} p(I) &= p(h_m)P(h_m|c)P(c_{m+1}|h) \prod_{n=m+2}^{N-1} P(h_n|h), \\ p(II) &= p(h_m)P(h_m|h)P(h_{m+1}|h) \prod_{n=m+2}^{N-1} P(h_n|h) \end{aligned} \tag{6.19}$$

and

$$\begin{aligned} q_I &= qs_m \sigma \left(\prod_{n=m+2}^{N} s_n \right) \sigma \\ q_{II} &= qs_m s_{m+1} \left(\prod_{n=m+2}^{N} s_n \right) \sigma \end{aligned} \tag{6.20}$$

Using eqns (6.19) and (6.20) in eqn (6.18) we have

$$\frac{P(h_m|c)P(c_{m+1}|h)}{P(h_m|h)P(h_{m+1}|h)} = \frac{\sigma}{s_{m+1}}. \quad (6.21)$$

In like manner consideration of the sequences

$$(I) \quad \ldots \overset{m}{c} \; h \; c \; c \; c \; c \; \overset{N}{c},$$
$$(II) \quad \ldots \overset{m}{c} \; c \; c \; c \; c \; c \; \overset{N}{c}, \quad (6.22)$$

leads to

$$\frac{P(c_m|h)P(h_{m+1}|c)}{P(c_m|c)P(c_{m+1}|c)} = \sigma s_{m+1}. \quad (6.23)$$

Using eqn (6.15), eqns (6.21) and (6.23) give the recursion relations

$$P(h_m|h) = \frac{1}{1 + \sigma P(h_{m+1}|h)/s_{m+1}P(c_{m+1}|h)},$$

$$P(c_m|c) = \frac{1}{1 + \sigma P(c_{m+1}|c)/s_{m+1}P(h_{m+1}|c)},$$

$$P(h_m|c) = 1 - P(h_m|h),$$

$$P(c_m|h) = 1 - P(c_m|c). \quad (6.24)$$

Starting with eqns (6.10) and (6.14), eqn (6.24) allows us to calculate all the conditional probabilities successively from $m = (N-1)$ to $m = 0$.

To calculate the *a priori* probabilities we again use a basic property of probabilities:

$$p(h_{m+1}) = p(h_m)P(h_m|h) + p(c_m)P(c_m|h). \quad (6.25)$$

Since

$$p(c_m) + p(h_m) = 1 \quad (6.26)$$

eqn (6.25) can be written as

$$p(h_{m+1}) = P(c_m|h) + p(h_m)\{P(h_m|h) - P(c_m|h)\}. \tag{6.27}$$

From eqn (6.24) all the conditional probabilities can be successively calculated. Hence if $p(h_1)$ is known, all the $p(h_m)$ can be successively calculated from eqn (6.27). To calculate $p(h_1)$ consider the two sequences

$$\begin{array}{lll} & 0 & N \\ (I) & c\ h\ h\ h\ \ldots\ h\ h, \\ & 0 & N \\ (II) & h\ h\ h\ h\ \ldots\ h\ h. \end{array} \tag{6.28}$$

The ratio of the partition functions for the two sequences above gives

$$\frac{p(I)}{p(II)} = \frac{p(c_1)P(c_1|h)}{p(h_1)P(h_1|h)} = \frac{q_I}{q_{II}} = \frac{1}{s_1} \tag{6.29}$$

or using eqn (6.26),

$$p(h_1) = \frac{1}{1 + P(h_1|h)/s_1 P(c_1|h)}, \tag{6.30}$$

and the problem of calculating the probability profile for a specific-sequence macromolecule in the nearest-neighbour model is solved.

The approach just outlined is extraordinarily simple to program for evaluation by a computer. To emphasize the simplicity of this method we schematically outline the approach:

Step 1: Calculate the conditional probabilities for $(N-1)$th unit (eqns (6.10) and (6.15))

Step 2: successively evaluate the conditional probabilities for the mth unit in the chain in terms of those for the $(m+1)$th unit starting with step 1 (eqn (6.24))

Step 3: calculate the *a priori* probabilities for the first unit in the chain using the conditional probabilities calculated in step 2 (eqn (6.30))

Step 4: successively evaluate the *a priori* probabilities for the mth unit in terms of the *a priori* probability of the $(m-1)$th unit and the condition probabilities already calculated starting with step 3 (eqn (6.27)). (6.31)

IN SPECIFIC-SEQUENCE MACROMOLECULES   311

Notice that the number of steps required is linear in the chain length N. On modern computers we can calculate probability profiles for molecules hundreds of units long in the nearest-neighbour model in a few seconds of computer time using the method of (6.31).

We now extend the above method to specific-sequence molecules, where unwinding can take place internally with loop formation or from the ends forming free chains.

6.2. Probability profiles; long-range correlations

The helices of DNA and collagen, being double-strand and triple-strand complexes respectively, can unwind internally forming loops or from the ends forming sequences of random coil with one end only fixed. As we have seen in Chapter 3, the statistical weight of a loop depends on the size of the loop and this size dependence is not a simple product of factors. Thus in order to correctly assign loop statistical weights we must know whether a coil sequence is on the end of the molecule (in which case it is a free chain and does not require a loop statistical weight, this being a very important distinction) or in the interior and how large the loop is. The requirement of having to know how large the loop is greatly complicates the problem since we must now correlate all the units in a loop. Since the largest loop has component chains that are almost the total length of the molecule (both ends pinched shut with a minimum number of helical units) we must essentially correlate all the units in the chain. It can be shown that to treat this problem using the matrix method requires $N \times N$ matrices, where N is the chain length. Thus to treat a DNA molecule with $N \sim 5000$ by using eqn (6.5) would require a specific-sequence product of 5000 (5000 × 5000) matrices.[4] This task is impossible to carry out even with the largest computers available today. There are reasonable approximations which reduce the problem to one requiring the order of 10 000 (100 × 100) matrices,[4] a manageable though cumbersome calculation. With the method of the previous section this calculation can be performed economically and exactly (no approximations made for the model assumed).[3]

In the last section where we treated the nearest-neighbour model, it was sufficient to describe the molecule in terms of the nearest-neighbour conditional probabilities which together with the *a priori* probabilities were sufficient to determine the *a priori* probability of any sequence of states. For example, the *a priori* probability of the sequence of states

$$\ldots x\ x\ h\ h\ c\ h\ c\ c\ x\ x \ldots$$

in the nearest-neighbour model is given by

$$p(hhchcc) = p(h)P(h|h)P(h|c)P(c|h)P(h|c)P(c|c).$$

For the case where interior coil sequences produce loops, we introduce conditional probabilities having a one-to-one correspondence with the nature of the statistical weights in the problem (specifically, how many units are correlated). Thus when loops are present the *a priori* probability of the sequence would be given by

$$p(h)P(h|h)P(h|3c|h)P(h|h)P(h|h)P(h|4c|h).$$

The probabilities required are specifically:

$p(h_m)$, $p(c_m)$ *a priori* probabilities that unit m is in the h or c state.

Conditional probability that if the mth unit is h:

$P(h_m|h)$ the $(m+1)$th unit is also h,

$P(h_m|nc|h)$ a sequence of n cs bordered by an h follows,   (6.32)

$P(hm|(N-m)c)$ a sequence of $(N-m)$cs follows to the right end of the chain

and

$P(c_1|nc)$ conditional probability that there is a sequence of n cs starting on the left end of the chain it is followed by an h at unit $(n+1)$.

The above probabilities have the conservation relations

$$P(h_m) + p(c_m) = 1,$$

$$P(h_m|h) + \sum_{n=1}^{n-1-m} P(h_m|n_c|h) + P(h_m|(N-m)c) = 1 \qquad (6.33)$$

$$\sum_{n=1}^{N} P(c_1|nc) = 1.$$

As with the nearest-neighbour model of the previous section our approach will be as outlined in eqn (6.31); successive calculation of conditional probabilities followed by successive calculation of *a priori* probabilities. First we show that all the conditional probabilities can be expressed in terms of $P(h_m|h)$.

The approach we use is the same as in the previous section: we consider specific sequences of states and construct conditional probabilities by using appropriate ratios of partition functions; the sequences chosen are such that the ratio of partition functions leaves only a few specific statistical weights. First consider the sequences

$$
\begin{array}{ll}
(I) & \overset{m}{\ldots h} \; c \; c \; c \; c \; c \; c \; c \; \overset{N}{c}, \\
(II) & \overset{m}{\ldots h} \; c \; c \; c \; c \; \overset{m+n}{h} \; h \; h \; \overset{N}{h}, \\
(III) & \overset{m}{\ldots h} \; h \; h \; h \; h \; h \; h \; h \; \overset{N}{h},
\end{array}
\qquad (6.34)
$$

which immediately give (following the procedure of the last section)

$$q_I/q_{II} = P(h_m|(N-m)c)/\prod_{k=m}^{N-1} P(h_k|h) = v_R(N-m),$$

$$q_{II}/q_{III} = P(h_m|nc|h)/\prod_{k=m}^{m+n} P(h_k|h) = v_I(m,n),$$

$$(6.35)$$

where

$v_R(N-m)$ = statistical weight of an end coil sequence $(N-m)$ units long on the right end of the molecule,

(6.36)

$v_I(m,n)$ = statistical weight of an interior sequence n units long bordered by h at units m and $(m+n+1)$.

Since the long-range correlations are present in sequences of coil states, for the present model we take the all-h state as the reference of free energy (statistical weight of all-h state thus being unity).

Solving eqn (6.35) for the conditional probabilities involving coil sequences gives

$$P(h_m | (N-m)c) = P(h_m|h) v_R(N-m) \sum_{k=m+1}^{N-1} P(h_k|h),$$

(6.37)

$$P(h_m | nc | h) = P(h_m|h) v_I(m,n) \sum_{k=m+1}^{m+n} P(h_k|h).$$

Using the conservation relations of eqn (6.33), eqn (6.37) gives

$$P(h_m|h) = \left[1 + \sum_{n=1}^{N-m-1} v_I(m,n) \prod_{k=m+1}^{m+n} P(h_k|h) + v_R(N-m) \prod_{k=m+1}^{N-1} P(h_k|h) \right]^{-1}$$

$(m = 1, N - 2)$

(6.38)

Eqn (6.38) is a recurrence relation for the $P(h_m|h)$; i.e., knowing $P(h_{N-1}|h)$ we can calculate $P(h_{N-2}|h)$; knowing $P(h_{N-1}|h)$ and $P(h_{N-2}|h)$ we can calculate $P(h_{N-3}|h)$, etc. The recurrence process is started by calculating $P(h_{N-1}|h)$. This quantity is calculated by using eqns (6.7) and (6.8). We obtain

$$P(h_{N-1}|h) = \left(1 + v_R(1) \right)^{-1}.$$

(6.39)

Eqn (6.38) gives all the $P(h_m|h)$, starting with eqn (6.39). Knowing all the $P(h_m|h)$ all the other conditional probabilities are given from eqn (6.37).

Having shown how to compute all the conditional probabilities we turn to the *a priori* probabilities. To do this we again consider specific sequences of states (states of the units shown fixed)

$$
\begin{array}{cccccc}
 & & m-2 & m-1 & m & m+1 \\
1 & & & . & . & h & \ldots \\
 & & . & . & h & h & \ldots \\
 & & . & h & c & h & \ldots \quad (6.40) \\
 & \ldots h & c & c & h & \ldots \\
c \ldots & \vdots & \ldots c & c & c & h & \ldots
\end{array}
$$

The *a priori* probability that unit $(m+1)$ is helix is given by the ratio of the partition functions

$$p(h_{m+1}) = q(\ldots h \ldots)/q$$

$$= \{q(\ldots hh\ldots) + q(\ldots hch\ldots) + q(\ldots hcch\ldots) + \ldots\}/q, \quad (6.41)$$

where the qs are the partition functions for the respective sequences shown in (6.40), q being the partition function for the molecule with no units fixed. Eqn (6.41) gives

$$p(h_{m+1}) = p(c_m)P(c_1|mc) +$$
$$+ \sum_{j=1}^{m-1} p(h_j)P(h_j|(m-j)c|h) + p(h_m)P(h_m|h), \quad (6.42)$$

which is a recurrence relation for $p(h_{m+1})$ in terms of $P(h_1|h)$ to $P(h_m|h)$ and $p(h_1)$ to $p(h_m)$.

The recurrence process of eqn (6.42) is started with $p(h_1)$. To obtain this quantity consider the sequences

$$
\begin{array}{lcccccccccccc}
 & 0 & & & & & & n+1 & & & & N & \\
(I) & c & c & c & c & c & c & h & h & h & h & \ldots & h & h, \\
 & 0 & & & & & & & & & & N & \\
(II) & h & h & h & h & h & . & . & . & . & . & . & h & h.
\end{array}
\quad (6.43)
$$

The ratio of the partition functions for the sequences in (6.43) gives

$$\frac{q_I}{q_{II}} = \frac{p(c_1)P(c_1|nc)}{p(h_1)\sum_{K=1}^{n}P(h_K|h)} = v_L(n), \qquad (6.44)$$

where $v_L(n)$ is the statistical weight of n coil units on the left end of the chain. Using eqn (6.33), eqn (6.44) gives

$$p(h_1) = \left(1 + \sum_{n=1}^{N-1} v_L(n) \prod_{k=1}^{n} P(h_k|h) + v_L(n) \prod_{k=1}^{N-1} P(h_k|h)\right)^{-1}. \qquad (6.45)$$

Using eqns (6.33) and (6.37), eqn (6.42) can be written as

$$p(h_{m+1}) = p(h_1)v_L(m)\prod_{k=1}^{m} P(h_k|h) +$$

$$+ \sum_{j=1}^{m-1} p(h_j)v_I(j,m-j) \prod_{k=j}^{m} P(h_k|h) + \qquad (6.46)$$

$$+ p(h_m)P(h_m|h).$$

Once again I shall outline schematically the approach to emphasize its simplicity.

    Step 1: calculate $P(h_{N-1}|h)$ (eqn (6.39)).
    Step 2: successively calculate all the $P(h_m|h)$ starting with step 1 (eqn (6.38)),
    Step 3: calculate $p(h_1)$ in terms of the $P(h_m|h)$ cal- (6.47) culated in step 2 (eqn (6.45)),
    Step 4: successively calculate all the $p(h_m)$ starting with step 3 and the (h h) from step 2 (eqn (6.46)).

The procedure of (6.47) gives the detailed probability profile. The average number of helix states is given by

$$\langle N_n \rangle = \sum_{m=1}^{N} p(h_m), \qquad (6.48)$$

while the average number of coil states is given by

IN SPECIFIC-SEQUENCE MACROMOLECULES   317

$$\langle N_c \rangle = 1 - \langle N_h \rangle \qquad (6.49)$$

Since interior unwinding is unfavoured becuase of the loop effect (entropy of a loop being less than that of a free chain), it is of interest to calculate the average degree of unwinding from each end. Defining

$p_m(\text{ch})_L$: *a priori* probability of a run of m coil units on the left end of the molecule,

$p_m(\text{hc})_R$: *a priori* probability of a run of $(N-m)$ coil units on the right end of the molecule,

then the average number of coil units on either end of the molecule is

$$\langle N_c \rangle_L = \sum_{m=1}^{N-1} m p_m(\text{ch})$$

$$= p(h_1) \sum_{m=1}^{N-1} m v_L(m) \prod_{k=1}^{m} P(h_k|h), \qquad (6.50)$$

$$\langle N_c \rangle_R = \sum_{m=1}^{N-1} (N-m) p_m(\text{hc})$$

$$= \sum_{m=1}^{N-1} (N-m) v_R(N-m) p(h_m) \prod_{k=m}^{N-1} P(h_k|h).$$

Using eqns (6.50) and (6.49) we have for the average number of interior (loop) coil units

$$\langle N_c \rangle_I = \langle N_c \rangle - \langle N_c \rangle_L - \langle N_c \rangle_R. \qquad (6.51)$$

Using the present method we can also treat the dissociation of the complex into single strands. We represent the dissociation equilibrium by the reaction

$$S_n \rightleftharpoons nS, \qquad (6.52)$$

S_n representing the n-strand complex ($n = 2$ for DNA, $n = 3$ for collagen) and S a single strand, with the conservation relation

$$[S] + n[S_n] = c_0, \tag{6.53}$$

c_0 being the total concentration of single strands. When the all-c state corresponds to chain dissociation, then the $v_L(N)$ term in eqn (6.45) must be deleted and the all-c state treated as dissociation. For the reaction of eqn (6.52) the equilibrium-constant expression is

$$\frac{[S]^n}{[S_n]} = \frac{n[S]^n}{c_0 - [S]} = \frac{v_L(N)}{\beta q}, \tag{6.54}$$

where β is the ratio of translational partition functions (see Chapter 3) representing the initial association of the strands, $v_L(N)$ is the statistical weight (excluding the translational entropy) of the all-c state (not equal to unity since we are taking the all-c state as the reference of free energy), and q is the partition function for the complex (again excluding the translational entropy, an effect included in β). It is of interest to note that in this whole discussion this is the first time the total partition function for the complex has been mentioned. Ordinarily, we first formulate the total partition function and then calculates average quantities by taking appropriate derivatives of the partition function. In the present approach we have first formulated appropriate average quantities (*a priori* and conditional probabilities) and only for chain dissociation do we need to evaluate the partition function. And we will show that the partition function q for the complex can be formulated in terms of the probabilities already calculated!

To calculate q we note that the probability of the all-helix state is given by

$$p(\text{all-h}) = \frac{1}{q} = p(h_1) \prod_{K=1}^{N-1} P(h_K|h), \tag{6.55}$$

where the statistical weight of the all-h state is unity since the all-h state has been chosen as the reference of free energy. Then

$$q = \left\{p(h_1) \prod_{K=1}^{N-1} P(h_K|h)\right\}^{-1}. \tag{6.56}$$

Defining

$$\zeta = [S]/c_0, \tag{6.57}$$

ζ being the fraction of dissociated single strands, and using eqn (6.56) for q, eqn (6.54) gives

$$\frac{n\zeta^n}{1-\zeta} = v_L(N)p(h_1) \prod_{k=1}^{N-1} P(h_k|h)/\beta c_0^{n-1}. \tag{6.58}$$

From (6.47) we have $p(h_1)$ and the $p(h_k|h)$; thus eqn (6.58) is a simple equation for ζ.

The quantity that is experimentally available is the net fraction of helix states in the system. This is given by the fraction of single strands in the associated form times the fraction of helix states in the complex. Then

$$\vartheta_h = (\text{fraction of complex}) \left(\frac{\langle N_n \rangle}{N}\right), \tag{6.59}$$

where

$$(\text{fraction of complex}) = 1 - \zeta \tag{6.60}$$

and $\langle N_n \rangle$ is given by eqn (6.49).

Finally the various contributions to the average fraction of coil states are given as

$$\vartheta_c = \vartheta_c(\text{single strands}) + \vartheta_c(\text{ends}) + \vartheta_c(\text{interior})$$

$$= \zeta + (1-\zeta)\left(\frac{\langle N_c \rangle_L}{N} + \frac{\langle N_c \rangle_R}{N}\right) + (1-\zeta)\frac{\langle N_c \rangle_I}{N}, \tag{6.61}$$

the quantities $\langle N_c \rangle_L$, $\langle N_c \rangle_R$, and $\langle N_c \rangle_I$ being given by eqns (6.50) and (6.51).

The specific form of the factors v_R, v_L, and v_I for nucleic acids is as follows. Since we are taking the all-h state as the zero of free energy, we assign each coil state the statistical weight

$$r_m = 1/s_m, \tag{6.62}$$

where s_m is the standard helix statistical weight (for DNA s_m would be either s_{AT} or s_{GC} depending on the position m in the sequence). Then introducing the loop statistical weight discussed in Chapter 3, the statistical weights required are

$$v_L(n) = \prod_{k=1}^{n} r_k,$$

$$v_I(m,n) = \sigma n^{-c} \prod_{k=m+1}^{m+n} r_k, \tag{6.63}$$

$$v_R(N-m) = \prod_{k=m+1}^{N} r_k.$$

With the form of eqn (6.63) it is convenient to define the variable

$$t_k = r_k P(h_k|h). \tag{6.64}$$

Using t_k of eqn (6.64), the previous relations have a very simple form. Since the substitution is straightforward I will not reproduce the results here.

We pause to reflect on what has been accomplished in this section. First we note that the problem being treated is unique in statistical mechanics in that we are discussing a system having units with different statistical weights, the order of these units being specified in detail. There are, of course, many examples of systems having different statistical weights for different units, e.g. copolymers, solutions, and alloys. But in the treatment of these systems, we average over all configurations or treat a specific simple arrangement (such as a copolymer with alternating units). Thus in this section a problem unique to biological macromolecules has been treated: the detailed probability profile of a model for a linear array of units having sequence-dependent statistical weights (the sequence is not a simple repeating pattern but a unique code) with long-range correlations, a difference between the ends and the interior of the molecule, and a concentration-dependent

IN SPECIFIC-SEQUENCE MACROMOLECULES   321

biomolecular nucleation has been treated exactly. With this rather complicated multiple equilibria it seems that we have come a long way from section 1.1, where we treated equilibrium in the simple two-level system. Yet if we go through this section and the last section of this book carefully, we will find that the final results are achieved by the repeated use of the same argument which is schematically as follows. Given the two states of a molecule, state (I) and state (II) then

$$p(\text{I relative to II}) = \frac{q_I}{q_I + q_{II}}, \qquad (6.65)$$

where the qs are the appropriate partition functions or statistical weights for the respective states. But eqn (6.65) is just a variant of eqn (1.3)! Thus the treatment of specific-sequence molecules ultimately reduces to consideration of component AB-type of reactions.

6.3. Nature of the equilibrium in specific-sequence molecules

DNA contains two types of unit in the double helix: A-T and G-C base pairs.[4] The helix stability parameters for these base pairs are

$$s_{AT} = \exp(\Delta h_{AT}(T - T_{AT})/RTT_{AT}),$$

$$s_{GC} = \exp(\Delta h_{GC}(T - T_{GC})/RTT_{GC}),$$

$$\Delta h_{AT} = -31.8 \text{ kJ mol}^{-1},$$

$$T_{AT} = 342.5 \text{ K} \qquad (6.66)$$

$$\Delta h_{GC} = -35.9 \text{ kJ mol}^{-1},$$

$$T_{GC} = 383.5 \text{ K}.$$

At about 363 K we have

$$\begin{aligned} s_{AT} &\sim \tfrac{1}{2} \\ s_{GC} &\sim 1 \\ s_{AT} s_{GC} &\sim 2. \end{aligned} \qquad (6.67)$$

Thus for a molecule containing approximately 50 per cent each
of A–T and G–C base pairs, the average helix statistical weight
will be unity at approximately 363 K and the molecule (for
long chains) will be approximately 50 per cent in the helix
state. Using the statistical weights of eqns (6.66) and (6.67)
as examples, in this section we investigate the nature of the
helix-coil equilibria on a molecule having a specific sequence
of two units.

As an example we treat a specific sequence 1000 unit long
have equal numbers of A–T and G–C pairs, generating the sequence by picking randomly A–T or G–C (this is equivalent to
generating the sequence by throwing a fair coin, heads = A–T,
tails G–C). A sample random sequence of 12 units is shown
below (for simplicity 0 = A–T, 1 = G–C; reclal that for random
occurrence, the average length of a run of 1s or 0s is 2)

sequence	1	0	0	1	1	0	0	0	1	0	1	1	
statistical weight of each unit	2	$\frac{1}{2}$	$\frac{1}{2}$	2	2	$\frac{1}{2}$	$\frac{1}{2}$	$\frac{1}{2}$	2	$\frac{1}{2}$	2	2	(6.68)
state favoured	h	c	c	h	h	c	c	c	h	c	h	h.	

In (6.68) I have illustrated the statistical weight each unit
would contribute to a helix and whether the helix or coil
state is favoured for each unit (taking the coil statistical
weight for both as unity). Thus on the basis of helix statistical weights alone we see that there are very short alternating
runs of strong helix formers (G–C units) and weak helix formers
(A–T units). The question is whether helix statistical weights
alone indeed strongly influence the occurrence of helix states
in the sequence. If the details of the occurrence of strong
helix formers in the sequence determine the probability of
helix states, then we would expect to find many very short
sequences of helix (average length 2 units) in our example of
1000 units with the specific sequence generated by random statistics.

Fig. 6.1 shows the probability profile for such a model
sequence at two temperatures in the neighbourhood of 363 K
taking σ = 10^{-4}. Two important features are immediately apparent

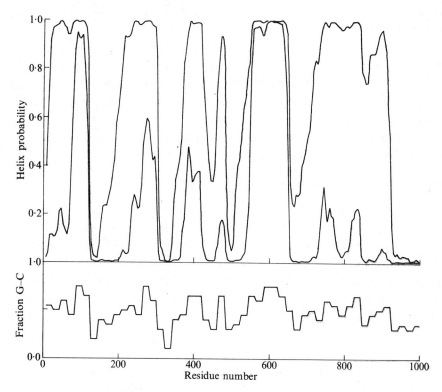

FIG. 6.1. Helix probability profiles at 360, 363, and 366 K for a random sequence having equal numbers of residue types having A–T and G–C statistical weights (eqn (6.65)); $\sigma = 10^{-4}$. The sequence is shown by the average fraction of G–C over blocks of 20 residues.

in Fig. 6.1. First the probability profile has a distinctly all-or-none character, there being regions where the probability of helix is ~100 per cent or ~0 per cent but very few regions having an intermediate probability; thus sequences of helix tend to be strongly localized in the molecule. Second, the length of well-characterized helix sequences is of the same order as we would expect for a homopolymer (see Chapter 4):

$$\langle L \rangle \sim \sigma^{-\frac{1}{2}} \pm \tfrac{1}{2}\sigma^{-\frac{1}{2}} \sim 100 \pm 50. \tag{6.69}$$

Since the average length of helix sequences (~100 units) is much longer than the average sequence length of strong helix formers (~2 units) it is clear that there is no correlation on a one-to-one basis between the *a priori* probability that a unit is in the helix state and the type of unit (strong or weak helix former). Since the helix regions are well defined in Fig. 6.1, the question arises, as to what defines these regions. In Fig. 6.1 the average fraction of G—C units in blocks 20 units long is plotted as a function of the position in the sequence. This quantity represents the tendency for there to be fluctuations from the average composition of 50 per cent G—C over blocks of units, the size of the block being much larger than a single unit and approaching $\langle L \rangle$. We see that this coarse-grained or blocked composition correlates well with the occurrence of helix regions. Thus the local details of the sequence do not determine helix or coil probabilities, but the average composition over a longer range determines the nature of the probability profile. Thus A—T units in a region of high G—C content are carried along with the G—C units to form helix; likewise, G—C units in a region of high A—T content are carried along with A—T units to form coil. To first approximation we can understand the structure of probability profiles by realizing that the average helix sequence length is of the same order of magnitude as it is in a homopolymer; unlike the case of the homopolymer, the sequence of helix does not diffuse up and down the length of the polymer but is localized at regions that are statistically high in strong helix former.

In Chapter 4 we showed that the average helix sequence length was determined by the competition between the tendency to reduce the number of σ factors (thus reducing the number of helix sequences, σ reflecting the unfavourable free energy of the ends of helix sequences) and to maximize the combinatorial entropy (these effects favouring respectively long and short sequences). Thus understanding $\langle L \rangle$ requires statistical-mechanical analysis and is essentially a statistical concept (it is important to realize that there is no potential energy operating over $\langle L \rangle$ units 'holding' them together). Since $\langle L \rangle$ persists in the probability profile for specific-sequence molecules, it is clear that statistics and not solely local energetics are very important in understanding conformational propensities (thus it is clear that conformational energy

IN SPECIFIC-SEQUENCE MACROMOLECULES   325

calculations cannot give the entire picture in understanding
specific-sequence molecules).

While the example we have given employed statistical weights
appropriate to DNA, it did not include the effect of loops.
Since no complete DNA sequence is yet known, I will not present
a sample DNA calculation here. Such calculations (including
the loop effect) based on model sequences are given elsewhere;[4]
the probability profiles are qualitatively the same as those
shown in Fig. 6.1. These model calculations also show that
helix sequences are well localized in the sequence and that
unwinding from the ends has about the same probability as unwinding from the interior (via loops) for chains 5000 units
long. I shall now give an illustration of the method of section
6.2 using the triple-strand helix of collagen as an example.

6.4. Probability of interior unwinding in collagen

In Chapter 3 I gave estimates for the statistical weights
for collagen. Eqn (3.116) (p. 221) gives the appropriate parameters and Table 3.4 shows s as a function of the number of
imino acids in a triplet unit. Compared with DNA parameters
(eqns (6.66) and (6.67)), the s values in Table 3.4 have a
broader range of variation. Thus one would expect that the
effect of specific sequence in collagen might be even more pronounced then in DNA (e.g. as in Fig. 6.1). In fact the probability profiles for collagen are quite different from those
for model DNA sequences, but not solely because of the difference in the range of variation of the appropriate s factors.[5]

For a loop complex involving m chains joined in common at
each end, each chain containing n units, the statistical weight
that reflects the loss in entropy caused by the constraint on
the chains, is σn^{-c}, where $c = \frac{3}{2}(m-1)$ if excluded volume is
ignored; for the double-strand DNA helix and the triple-strand
collagen helix $c = \frac{3}{2}$ and $c = 3$ respectively. Estimates of the
effect of excluded volume (see Chapter 3) for DNA-like loops
$c = 1.75$; by analogy a better estimate for collagen would be
$c \sim 3.5$. Consider then the effect of introducing an interior
run of coil states in the triple-strand helix of collagen producing a three-strand loop. If there are n triplets in each

component chain in the loop, then the loop factor is $n^{-3.5}$, deleting the σ factor. If σ is 10, then this factor is $10^{-3.5}$; since we would expect $\sigma \ll 1$, the introduction of σ makes the loop factor even smaller. Thus we easily see that interior coil states are very improbable in collagen.

For collagen model sequences of 300 triplets (in each strand) or 900 residues, there is essentially no tendency for unwinding to take place via internal loops. Since in Chapter 3 I showed that known collagen sequences follow random statistics as far as the placement of imino acids is concerned, model collagen specific sequences generated by random placement of imino acids have been examined allowing many sequences to be studied. Fig. 6.2 shows the probability profiles for one such sequence at two temperatures using the parameters of Table 3.4

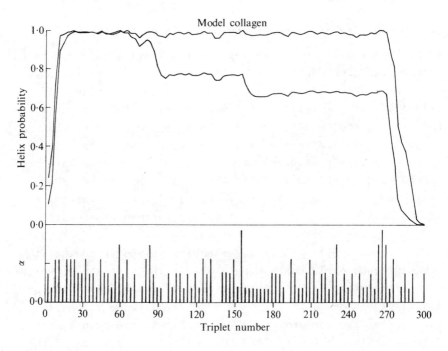

FIG. 6.2. Helix probability profile for a model collagen sequence (random placement of imino acids). The sequence is shown as the average fraction, α, of imino acids in successive groups of three triplets; the net average is $\alpha = 0.220$. Interior loops are allowed as is unwinding from both ends.[5]

where the sequence is shown in terms of the fraction α of imino acids in successive groups of three triplets. We see that there is no significant tendency for distinct interior coil regions to form. Rather, all unwinding is from the ends, proceeding in well-defined plateaux. Note that in this example essentially all the unwinding takes place from the right end. In natural collagen one end of the molecule is covalently tied together. To see if interior loops will be favoured with one end closed, a calculation may be performed using the same sequence shown in Fig. 6.2 choosing to close the right end, with the proviso that a loop is already nucleated near the covalently bound end (the ends of the collagen sequence do not show the Gly-x-y structure and thus it is reasonable to assume that the ends do not form triple helix). Thus the model calculation favours interior loop formation: from Fig. 6.2 the right end is the weaker helix former; and the right end is closed with a loop already nucleated. Fig. 6.3 shows the results of the calculation at several temperatures. We see that all the unwinding is now from the left end with no tendency for loop formation. Since the value of σ for collagen is unknown we chose the value $\sigma = 10^{-2}$ (loop formation from the right end in the model of Fig. 6.3 is independent of σ since we assumed a loop was already nucleated at the right end). From our model calculations it appears that collagen always unwinds from the ends; since loops are not present it is of course not possible to measure σ experimentally (unless we could synthesize a triple-strand complex covalently bound at both ends).

6.5. Probability profiles for globular proteins

With the breaking of the genetic code there remains a fascinating coding problem in molecular biology. And that is the problem of being able to infer the total conformation of a globular protein from the amino-acid sequence. To be able to predict the conformation of a protein with the same accuracy as is now obtained from X-ray studies is probably impossible. In this section we want to show, however, that even on a very crude level, we can say something about the conformational propensities of a protein based solely on the amino-acid

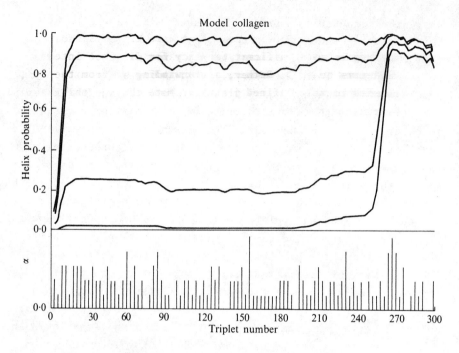

FIG. 6.3. Helix probability profile for the same model collagen sequence used in Fig. 6.2 with the right end closed and with a loop always nucleated at that end.[5]

sequence using helix probability profiles.[6]

To calculate the probability profiles for proteins we must have σ and s values for all 20 natural amino acids. Since we want the σ and s values appropriate to aqueous solution, these parameters must be determined from a study of the appropriate polymers in water. Unfortunately none of the homopolymers of the natural amino acids is water-soluble (except polyproline which does not form α-helix). This difficulty has been circumvented by two techniques, namely the use of sandwich compounds and random copolymers. In the sandwich-compound technique[7] a block of the desired homopolymer is flanked by blocks of a water-soluble group (e.g. D,L-lysine, the random mixture of D and L prohibiting a regular crystal structure hence enhancing solubility; also the sidechains are charged at pH ~ 7). In the random-copolymer technique,[8] a helix-forming synthetic polyamino acid that is water-soluble is synthesized with a certain fraction of a natural amino acid present as a copolymer unit. The natural amino acid acts as a perturbation on the helix coil transition of the parent molecule, allowing s and σ for the natural amino acid to be determined.

Table 6.1 shows the parameters that have so far been determined by the techniques outlined above for the natural amino acids. We see that the variation in s for the amino acids is small compared with the range of s values for DNA and collagen. However, note that there is a significant spread in the value of σ factors. Table 6.1 also shows $\langle L \rangle = \sigma^{-\frac{1}{2}}$ at $s = 1$; the fact that this quantity varies greatly is significant, since we have seen in section 6.3 that the value of σ largely determines the length of helical sequences even in specific-sequence molecules. From Table 6.1, a helix sequence nucleated by Leu has a much larger nucleation parameter σ than a helix sequence nucleated by Gly. Since the parameters for all the amino acids are not known, in order to do model calculations we must assign approximate parameters to all the amino acids based on their similarity to the amino acids given in Table 6.1. Such a classification is shown in Table 6.2; this is a very approximate procedure, which can be improved when more data are available. Using Tables 6.1 and 6.2 I shall now give some model calculations.

We use as an example sperm-whale myoglobin. This protein consists of 153 residues in a single chain with no disulphide bridges. The X-ray structure of this molecule is known in detail; in particular the α-helix regions in the conformation present in the crystal are known.[14] Using the technique of section 6.1 we give several calculations of the helix probability profile for myoglobin.[15] To illustrate the influence of heterogeneity of σ and s separately we calculate the probability profile using: (I) a homogeneous average s and σ heterogeneous; (II) a homogeneous average σ with s heterogeneous; and (III), both σ and s heterogeneous. Case I is shown in Fig. 6.4; we see that there is no tendency for distinct regions of helix and coil to form, there being little difference from the case of both σ and s homogeneous (shown by the smooth curve) other than a small fine structure. In assigning σ factors to a helix sequence beginning and ending with residues i and j respectively

TABLE 6.1

Parameters σ and s for various polyamino acids in water at 335 K.[†]

Amino acid	σ	$\sigma^{-\frac{1}{2}}$	s
Gly[9‡]	1.0×10^{-5}	310	0.63
Ser[10]	7.5×10^{-5}	115	0.74
PHPG[8§]	2.2×10^{-4}	68	0.96
Ala[11]	8.0×10^{-4}	35	1.01
Phe[12]	1.8×10^{-3}	24	1.00
Leu[13]	3.3×10^{-3}	18	1.09
Leu[7c]			1.3[¶]

[†]The σ and s values vary little with temperature; typical parameters are those for Phe: $\Delta h = -711$ J mol^{-1}; $T_0 = 370$ K.

[‡]See Notes and References.

[§]Polyhydroxypropylglutamine; a synthetic water-soluble polyamino acid.

[¶]Value obtained from the study of Leu sandwich compounds; the difference from the preceding value determined from random copolymers indicates that Leu in a Leu environment is a stronger helix former than in a random copolymer.

TABLE 6.2

Assignments of amino acids to classes according to similarity with the compounds listed in Table 6.1

Class	Parent compound	Compounds assigned
1	Gly	Pro
2	Ser	Asn
3	PHPG	Glu, Gln, Lys, Arg, Thr, Tyr, His, Trp, Met
4	Ala	
5	Phe	
6	Leu	Ileu, Val
7	Leu (followed by Leu)	Val or Ileu followed by Val, Ileu, or Leu

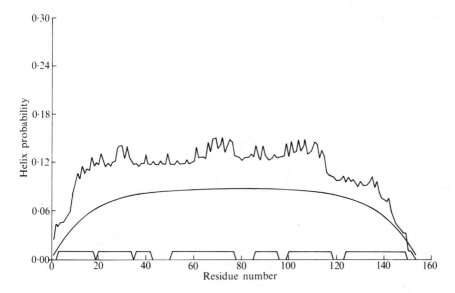

FIG. 6.4. Helix probability profile for myoglobin with s homogeneous, σ heterogeneous; see text for description of parameters. The smooth curve is for s and σ homogeneous; α-helix regions found in the crystal structure are shown by the plateaux.[15]

we use the approximation $\sigma_{ij} = \sqrt{(\sigma_i \sigma_j)}$; this is clearly a crude first approximation. Fig. 6.5 shows case II; there is now more of a tendency for there to be ups and downs in helix probability, but still there are no sharply delineated helix regions. In case III (Fig. 6.6), with σ and s both heterogeneous, there are now fairly well-defined regions of high and low helix probability.

At this stage in the calculations we can only note that based on amino-acid sequence along and the data of Table 6.1 (with the classification of Table 6.2) there are regions in the sequence that differ markedly in the tendency to form helix, these regions having a rough correlation with the regions that are found to be helix in the conformation found in the crystal (indicated by the plateaux at the bottom of the graphs). Several points should be kept in mind in considering these probability profiles. One is that the conformation of a globular protein is not determined solely by the helix-coil equilibrium (hence by σ and s) but rather is influenced by long-range

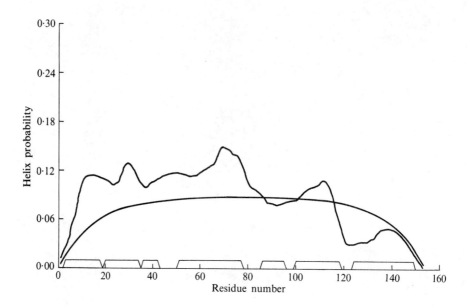

FIG. 6.5. Helix probability profile for myoglobin with s heterogeneous, σ homogeneous.[15] See caption to Fig. 6.4.

interactions (the chain folding back on itself forming a more-or-less globular structure). Hence the information provided by probability profiles is only a first step in identifying conformational preferences of regions and hence reducing the number of conformations to be considered in the attempt to determine the final conformation. Another point to keep in mind is that there is not necessarily a single conformation for a protein in solution and the conformation in solution need not be the same as the conformation found in the crystal.

The calculations shown in Figs. 6.4 to 6.6 take only seconds of computer time. It seems evident that such modest investments of time offer at least a promising beginning to the task of 'reading' the amino-acid sequence and promise better results when more data are available. These results, coupled with conformational energy calculations, imply that we can be optimistic that an understanding of at least the groww features if not the details of protein structure in terms of the concepts of physical chemistry is something that can be achieved.

The above calculations were based on the pseudo-nearest-

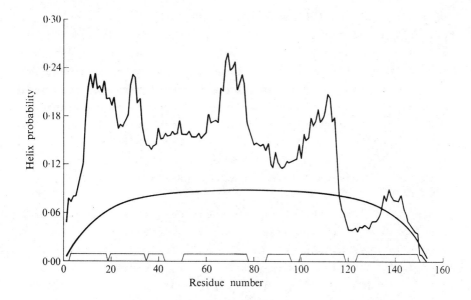

FIG. 6.6. Helix probability profile for myoglobin with s and σ heterogeneous.[15] See caption to Fig. 6.4.

neighbour model outlined in Chapter 4. I shall now introduce another sequence-dependent effect not present in the calculations so far, namely the electrostatic interaction between charged side-chains.[15] In the crystal structure of myoglobin there are four main coil regions. Numbering residues from the N terminus to the C terminus, the amino-acid sequences of these coil regions are:

	residues	sequences							
I	43–50	Phe	Asp −	Arg +	Phe	Lys +	His +	Leu	Lys +
II	78–85	Lys +	Lys +	Gly	His +	His +	Glu −	Ala	Glu −
III	119–124	His +	Pro	Gly	Asn	Phe	Gly		
IV	149–153	Leu	Gly	Tyr	Glu −	Gly.			

(6.70)

Regions III and IV contain several helix-breaking (low s values;

see Tables 6.1 and 6.2) amino-acid residues; in addition region IV is at the end of the chain and chain ends tend to be coil even in homopolymers (the smooth curve in Figs. 6.4 to 6.6 is for a homopolymer all units having the same $\langle s \rangle$ and $\langle \sigma \rangle$, the average being over the residues in myoglobin). In (6.70) the charge on the residue side-chains at pH $\sim$ 7 is indicated. We note that in both regions I and II there are high concentrations of positive charge. Since the distance between side-chains is greater when the backbone is coil than when it is helix, the repulsion between like charges is reduced if these regions are coil.

Table 6.3 lists the charge on the side-chains in myoglobin as a function of the position in the sequence. The negative charges are either Glu or Asp: the positive charges are Lys or Arg; the positive charges shown as (+) are His (to be discussed shortly), see Fig. 1.6 (p. 24) for the titration curves of the amine and carboxyl groups. In the α-helix the side-chain of residue i is brought in close proximity to the side-chains of residues $(i + 3)$ and $(i + 4)$. Thus, for example, the sequence

$$\begin{array}{ccc} + & - & - \\ \text{Lys} \ldots & \text{Glu} & \text{Asp} \\ 56 & 59 & 60 \end{array}$$

will produce a strong favourable interaction between Lys and both Glu and Asp (in the α-helix neighbouring charges are relatively far apart).

From the above considerations it seems clear that simple electrostatic interactions between side-chains could greatly influence helix probability profiles. To explore this effect the probability profile was calculated for myoglobin using the matrix technique of eqns (6.4) and (6.5); a 16 × 16 matrix was used correlating the ith residue with residues as far away as $(i + 4)$. The pseudo-nearest-neighbour parameters of Tables 6.1 and 6.2 were used together with an electrostatic statistical weight of the form

$$v = \exp(-\varphi(r)/RT), \qquad (6.70)$$

with $\varphi(r)$ given by the Debye-Hückel expression of eqn (2.258).

TABLE 6.3

Side-chain charges as a function of sequence position
(the symbol (+) indicates histidine)

Myoglobin

4	−	52	−	98	(+)
6	−	54	−	102	+
12	(+)	56	+	105	−
16	+	59	−	109	−
18	−	60	−	113	(+)
20	−	62	+	116	(+)
24	(+)	63	+	118	+
27	−	64	(+)	119	(+)
31	(+)	77	+	126	−
34	(+)	78	+	133	+
36	(+)	79	+	136	−
38	−	81	(+)	139	+
41	−	82	(+)	140	+
42	+	83	−	145	+
44	−	85	−	147	+
45	+	87	+	148	−
47	+	93	(+)		
48	(+)	96	+		
50	+	97	(+)		

Carboxypeptidase A

2	+	88	−	181	−
13	(+)	101	−	184	+
16	−	104	−	186	(+)
17	−	108	−	190	(+)
20	−	120	(+)	196	(+)
23	−	122	−	215	−
29	(+)	124	+	216	+
31	−	127	+	218	−
35	+	128	+	224	+
40	+	130	+	231	+
43	−	142	−	239	+
45	+	145	+	256	−
51	+	148	−	264	+
59	+	153	+	270	−
65	−	163	−	272	+
69	(+)	166	(+)	273	−
71	+	168	+	276	+
72	−	173	−	302	−
84	+	175	−	304	(+)
85	+	177	+		

An approximate set of appropriate distances was estimated from molecular models; the values used are shown in Table 6.4.

TABLE 6.4

Approximate distances between charged side-chains

Interaction between side-chain i and	Distance (nm) in helix	in coil
$i + 1$	1.00	1.00
$i + 2$	1.20	1.20
$i + 3$	0.79	1.00
$i + 4$	0.75	1.60

Fig. 6.7 shows the results of the above calculation for myoglobin at the four salt concentrations (1-1 electrolyte) of 0.0, 0.1, 1.0, and ∞ mol l^{-1}; as the salt concentration decreases, the probability of helix increases in myoglobin, the

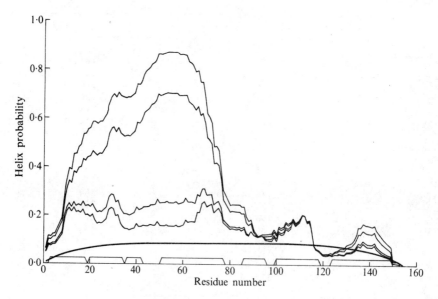

FIG. 6.7. Helix probability profile for myoglobin with σ and s heterogeneous including electrostatic interactions between side-chains; the curves from top to bottom are respectively for the salt concentrations 0.0, 0.1, 1.0, and ∞ mol l^{-1}.[15]

IN SPECIFIC-SEQUENCE MACROMOLECULES   337

curve for infinite salt concentration being the case of no
electrostatic interactions (complete shielding); the lowest
curve in Fig. 6.7 is identical with the curve shown in Fig.
6.6, although the scale is different. Clearly electrostatic
interactions produce a major effect, although the delineation
between helix and coil regions is not made more distinct. Since
the introduction of heterogeneity in σ (an effect included in
Fig. 6.7) as illustrated in comparing Fig. 6.5 (s heterogeneous, σ homogeneous) with Fig. 6.6 (s heterogeneous, σ heterogeneous) introduces major changes in the helix probability profile, and since the assignment of heterogeneous σs factors is
the least certain aspect of the present calculation (through
the assignment of a σ to a helix sequence of the form $\sqrt{(\sigma_i \sigma_j)}$),
it is of interest to see how much the large effect of electro-
static interactions indicated in Fig. 6.7 is influenced by our
choice of assigning σ factors. Fig. 6.8 shows the probability
profile including electrostatic interactions with heterogeneous
σ factors and with a homogeneous, average σ (smoother curve),
both at zero salt concentration; again the profile of Fig. 6.6

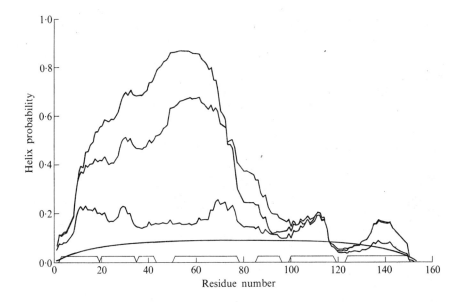

FIG. 6.8. Fig. 6.7 repeated for zero salt concentration (top curve) plus
the case (middle curve) where the histidines are uncharged.[15]

is reproduced for comparison. We see that the effect of electrostatic interactions remains largely unchanged.

From Fig. 1.6 it is clear that Glu and Asp ($pK \sim 4$) and Lys and Arg ($pK \sim 12$) will be (-) and (+) charged respectively in the range pH = 7 ± 1. On the other hand, His has a pK in the range 6-7. Hence this side-chain can be charged or uncharged in the region of pH ~ 7 (the state of charge being influenced by the state of charge of neighbouring residues). Fig. 6.9 shows a repeat of the electrostatic calculation with His (the (+) in Table 6.3 are His) in the uncharged form (middle curve); Fig. 6.7 (zero salt concentration) is repeated for comparison. The effect of discharging the His residues is seen to be large; it is thus clear that in molecules containing large numbers of His residues, this amino acid should be allowed to be in either the charged or uncharged form. Unfortunately this greatly increases the matrix size required to treat electrostatic effects.

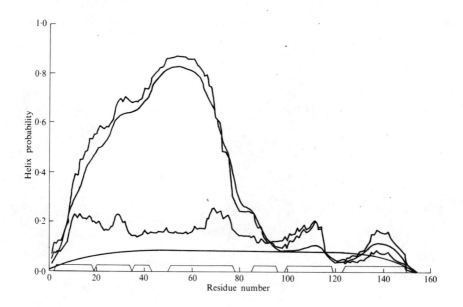

FIG. 6.9. Fig. 6.7 repeated for zero salt concentration plus the case (smoother curve) where σ is homogeneous.[15]

Myoglobin contains a high percentage of α-helix (~80 per cent) and also a large number of residues with charged side-chains. For comparison we shall treat the protein carboxypeptidase A. The amino-acid sequence and crystal structure of the protein are known.[16] It contains 307 residues and is about 38 per cent α-helix. Using the same approximations and parameters as used for the calculations on myoglobin, the probability profile for carboxypeptidase A is shown in Fig. 6.10. The regions found to be α-helix in the crystal are shown by the plateaux at the bottom of the graph; the smooth curve shows the probability profile using the same $\langle s \rangle$ and $\langle \sigma \rangle$ for each residue (averaging over the residues of carboxypeptidase A). The remaining curves show the probability profiles obtained using heterogeneous s and σ factors with (upper curve) and without electrostatic effects (at zero salt concentration). In this protein electrostatic effects are seen not to have a marked influence on the probability profile; the charges as a function of position in the sequence for this protein are

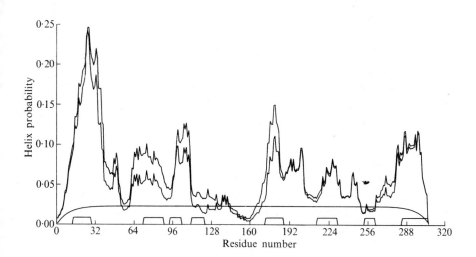

FIG. 6.10. Helix probability profiles for carboxypeptidase A: the upper curve includes electrostatic intersctions between side-chains with σ and s heterogeneous; the middle curve ignores electrostatic interactions but includes σ and s heterogeneous; the smooth curve is for σ and s homogeneous. The plateaux are α-helix regions found in the crystal structure.[15]

shown in Table 6.3. There are, however, distinct regions of high and low helix probability that correlate roughly with the position of α-helix found in the crystal.

6.6. Summary

In this chapter I have outlined techniques for treating multiple cooperative equilibria in specific-sequence biological macromolecules. These are particularly complicated equilibria involving literally billions of different species (an N-residue chain with only two states per residue has 2^N different molecular states), each state requiring a specific statistical weight or equilibrium constant that reflects the specific sequence of residue types and the particular sequence of residue states. It is obvious that this problem can be handled only with the systematic partition-function approach. Thus statistical mechanics is a necessary tool for the understanding of conformational transitions in biological macromolecules. And it is important to bear in mind that the main influences on equilibrium cannot be understood in terms of intermolecular forces alone; e.g. the correlation length, a very important concept in understanding probability profiles, whilst being a function of intermolecular forces, arises only in the framework of the statistical—mechanical analysis. Finally we note that the structure of the probability profiles is the net resultant of many competing effects (heterogeneity of s, of σ, and electrostatic effects) the final position of the equilibrium at any particular residue being influenced by the nature of the residues over a long range (many 10s of residues) on either side of the residue in question. Thus simple rules for interpreting the influence of sequences on conformation should not be expected necessarily to work well (e.g. in nucleic acids there is no correlation between whether a base pair is A—T or G—C and whether it is in the helix state — the correlation is with regions rich or poor in G—C). What can be systematically assigned to each residue are statistical weights (depending at most on the nature of a few nearest-neighbour residues); this set of statistical weights, incorporated into the framework of the appropriate statistical—mechanical analysis, then yields useful information.

Notes and References

1. See P&S, Chapter 9, section P.
2. Lacombe, R.H. and Simha, R., *J. chem. Phys.* **58**, 1043 (1973); Simha, R. and Lacombe, R.H., *J. chem. Phys.* **55**, 2936 (1971).
3. Poland, D., *Biopolymers* **13**, 1859 (1974).
4. P&S, Chapter 9.
5. Schwarz, M. and Poland, D., *Biopolymers* **13**, 1873 (1974).
6. For examples of helix probability profiles using a limited set of σ and s data see the following (note that for many of the profiles in the following references most of the structure of the profiles is determined by the assignment of the very low value $s = 0.385$ to Pro, Ser, Gly, and Asn (see Tables 6.1 and 6.2)). Lewis, P.N. and Scherage, H.A., *Archs Biochem. Biophys.* **144**, 526, 584 (1971).
7. (a) Gratzer, W.B. and Doty, P., *J. Am. chem. Soc.* **85**, 1193 (1963); (b) Ingwall, R.T., Scheraga, H.A., Lotan, N., Berger, A., and Katchalski, E., *Biopolymers* **6**, 331 (1968); (c) Ostroy, S.E., Lotan, N., Ingwall, R.T., and Scheraga, H.A., *Biopolymers* **9**, 749 (1970).
8. von Dreele, P.H., Lotan, N., Ananthanarayanan, V.S., Andreatta, R.H., Poland, D., and Scheraga, H.A., *Macromolecules* **4**, 408 (1971).
9. Ananthanarayanan, V.S., Andreatta, R.H., Poland, D., and Scheraga, H.A., *Macromolecules* **4**, 417 (1971).
10. Hughes, L.J., Andreatta, R.H., and Scheraga, H.A., *Macromolecules* **5**, 187 (1972).
11. Platzer, K.E.B., Ananthanarayanan, V.S., Andreatta, R.H., and Scheraga, H.A., *Macromolecules* **5**, 177 (1972).
12. Van Wart, H.E., Taylor, G.T., and Scheraga, H.A., *Macromolecules* **6**, 266 (1973).
13. Alter, J.E., Taylor, G.T., and Scheraga, H.A., *Macromolecules* **5**, 739 (1972).
14. For the sequence and structure of many proteins see: Dickerson, R.E. and Geis, I., *The structure and Action of Proteins*, Harper and Row, New York (1969).
15. Poland, D., unpublished calculations.
16. Quiocho, F.A. and Lipscomb, W.N., *Adv. Protein Chem.* **25**, 1 (1971).

APPENDIX: A SHORT REVIEW OF MATRIX ALGEBRA

In Chapter 4 we expressed partition functions in terms of a matrix product. In this Appendix I shall review some of the elementary notions of matrix algebra.

A matrix is an $M \times N$ array of numbers. Since in the applications we use the matrices are square ($M = N$) we shall limit our discussion to that case. We denote a matrix by any of the following forms

$$\mathbf{A} = \begin{bmatrix} a_{11} & a_{12} & \cdots & a_{1m} \\ a_{21} & a_{22} & \cdots & a_{2m} \\ \cdots & & & \\ \cdots & & & \\ a_{M1} & a_{M2} & \cdots & a_{MM} \end{bmatrix} = (a_{ij}). \tag{A.1}$$

The operation of multiplying the matrix $\mathbf{A}$ by the matrix $\mathbf{B}$ to give the matrix $\mathbf{C}$ is given by

$$\mathbf{C} = \mathbf{AB} = \left(\sum_{k=1}^{M} a_{ij} b_{jk} \right) \tag{A.2}$$

The quantity of interest in connection with evaluating partition functions is the product of matrices $\mathbf{W}^N$. This quantity can be evaluated by repeated application of eqn (A.2). However, a simpler method exists. If we have a matrix where all the elements are zero except those along the diagonal, e.g.

$$\mathbf{\Lambda} = \begin{bmatrix} \lambda_1 & 0 & 0 & 0 \\ 0 & \lambda_2 & 0 & 0 \\ 0 & 0 & \lambda_3 & 0 \\ 0 & 0 & 0 & \lambda_4 \end{bmatrix} \tag{A.3}$$

Then by applying eqn (A.2) we have

$$\mathbf{\Lambda}^N = \begin{bmatrix} \lambda_1^N & 0 & 0 & 0 \\ 0 & \lambda_2^N & 0 & 0 \\ 0 & 0 & \lambda_3^N & 0 \\ 0 & 0 & 0 & \lambda_4^N \end{bmatrix} \quad (A.4)$$

In matrix algebra it is shown that a matrix **W** is related to a diagonal matrix **Λ** by the following equation

$$\mathbf{W} = \mathbf{T\Lambda T}^{-1}, \quad (A.5)$$

where $\mathbf{T}^{-1}$ is the inverse of **T**, with

$$\mathbf{TT}^{-1} = \mathbf{I} = \mathbf{T}^{-1}\mathbf{T}. \quad (A.6)$$

The quantity **I** is the identity matrix, which is a diagonal matrix, the diagonal elements all of which are unity. The relation (A.5) holds for all real, symmetric ($a_{ij} = a_{ji}$) matrices. In general, eqn (A.5) also holds for matrices representing physical models even though the matrices are not symmetric. Using eqns (A.5) and (A.6) we have

$$\mathbf{W}^N = (\mathbf{T\Lambda T}^{-1})^N = \mathbf{T\Lambda}^N \mathbf{T}^{-1}. \quad (A.7)$$

Utilizing eqn (A.4), we see that the amount of matrix multiplication has been enormously reduced by introducing the diagonal matrix **Λ**. The problem remains of calculating the elements of **Λ** and **T**.

The λ_i, known as the eigenvalues of **W**, are obtained as the roots of the equation

$$|\mathbf{W} - \lambda \mathbf{I}| = \mathbf{0}. \quad (A.8)$$

If **W** is an $M \times M$ matrix, then expansion of the determinant in eqn (A.8) results in an Mth-order polynomial in λ yielding M roots. From eqns (A.5) and (A.6) we obtain equations for **T** and $\mathbf{T}^{-1}$

$$WT = T\Lambda, \qquad (A.9)$$

$$T^{-1}W = \Lambda T^{-1}.$$

More detail along with specific examples can be found in P&S, Appendix A.